Innovation and Technology Enhancing Mathematics Education

Perspectives in the Digital Era

MATHEMATICS EDUCATION IN THE DIGITAL ERA
Volume 9

More information about this series at http://www.springer.com/series/10170

Eleonora Faggiano · Francesca Ferrara
Antonella Montone
Editors

Innovation and Technology Enhancing Mathematics Education

Perspectives in the Digital Era

Springer

Editors
Eleonora Faggiano
Università di Bari Aldo Moro
Bari
Italy

Antonella Montone
Università di Bari Aldo Moro
Bari
Italy

Francesca Ferrara
Università di Torino
Torino
Italy

ISSN 2211-8136 ISSN 2211-8144 (electronic)
Mathematics Education in the Digital Era
ISBN 978-3-319-61487-8 ISBN 978-3-319-61488-5 (eBook)
DOI 10.1007/978-3-319-61488-5

Library of Congress Control Number: 2017944210

Printed on acid-free paper

This Springer imprint is published by Springer Nature
The registered company is Springer International Publishing AG
The registered company address is: Gewerbestrasse 11, 6330 Cham, Switzerland

Foreword by Ferdinando Arzarello

The issue of a fresh and creative use of technology to enhance innovation in mathematics education is a hot spot in current debates on mathematics education. Many countries invest a lot in equipping the schools with (more or less) updated devices and in organizing consequent teachers' training for a suitable use of the new tools.

The research is cautious in claiming that technology has great positive associations with educational outcomes (see, e.g. the Report of Higgins, Xiao, & Katsipataki, 2012). What is there underlined more is that:

> The range of impact identified in these studies suggests that it is not whether technology is used (or not) which makes the difference, but how well the technology is used to support teaching and learning. There is no doubt that technology engages and motivates young people. However, this benefit is only an advantage for learning if the activity is effectively aligned with what is to be learned. It is therefore the pedagogy of the application of technology in the classroom which is important: the how rather than the what. This is the crucial lesson emerging from the research. (p. 3)

Hence, it is this lesson that must be considered by researchers and practitioners: it underlines the necessity of approaching technology in the classroom from a wider standpoint, namely considering what Mishra and Koehler (2006) call the technological pedagogical content knowledge (*TPACK-perspective*).

Another issue about the type of impact that technology can have in schools emerges from PISA surveys. In one of the last PISA in Focus (n.64), it is pinpointed that

> even when most students have easy access to new media, inequalities persist in the way they use these tools. The use of online media depends on the student's own level of skills, motivation, and support from family, friends and teachers, which vary across socio-economic groups. In their free time, disadvantaged students tend to prefer chatting rather than sending e-mails. They are also much less likely to read the news or obtain practical information from the Internet, perhaps because their navigation and reading skills are often more limited than those of advantaged students. (p. 4)

In fact, PISA results show that proficiency in the ability to use ICT tools for learning is strongly related to more traditional school abilities:

> Proficiency in online reading and navigation requires students to plan and execute a search, evaluate the usefulness of information, and assess the credibility of sources on line – skills that schools can encourage students to practice and develop. [...] Proficiency in online reading and navigation requires students to plan and execute a search, evaluate the usefulness of information, and assess the credibility of sources on line. [...] students with good reading skills, regardless of their background, have a much easier time ending their way around–and mining the considerable assets of–the Internet.

The lesson here is that the activities with ICT should be designed according to a global standpoint of the teaching design for the classroom activities: let us call this the *global skills perspective*.

These two combined perspectives require that researchers deeply rethink the theoretical and empirical frames at the base of the educational projects for enhancing and improving mathematics teaching and learning in the classrooms. What is needed is not a cumulative programme where new devices are at stake together with the old ones focussing on possible hoped advantages for teachers' and students' activities, perhaps without any founded assumption. There is the necessity of a deeper insight, which touches the real roots of learning according the hewn findings that research puts forward not only from the pedagogical and technological innovation standpoint but also considering the new results of other disciplines, from neuro- to social sciences, which can give fresh ideas and programmes to pursue global learning and teaching designs, aligned with the two perspectives pointed out above.

In this sense, the book is very useful. From the one side, it offers some interesting suggestions for these new spaces for research and for innovation, pushing forward possible programmes of innovation linked to the last findings in technology: from the affordances allowed by touch screen devices to those that Wii-environments offer, and others. The interest of these proposals consists in the deep analysis of the intertwining between the cognitive, embodied and didactical affordances that such devices allow. From the other side, also more or less standard examples are considered and innovative uses of digital technologies are exemplified in different contexts: CAS environments, interactions between concrete and simulated artefacts, construction of mathematics concepts within institutional infrastructures, the use of a single computer in a classroom, the use of technology for students having mathematical learning difficulties or disabilities.

Overall, the chapters offer an interesting updated survey of important researches in the field, as pointed out in the retrospective Chap. "From Acorns to Oak Trees: Charting Innovation Within Technology in Mathematics Education": there, it is shown how the progress of innovations in this field has "been seeded and taken root" within the ICTMT community in the years.

In most of the book, the two perspectives—TPACK and global-skill one—are both present: therefore, the book can be read with benefits not only by researchers but also by practitioners.

Practitioners will find new ideas about an old issue that new technology today puts forward, but that is connected to older problems raised many centuries ago by a philosopher like Bacon (1620), who, at the beginning of the scientific revolution, summarized a main issue that even today teachers (and not only they) face when using artefacts, and specifically technological devices in their classrooms:

> Neither the naked hand nor the understanding left to itself can effect much. It is by instruments and helps that the work is done, which are as much wanted for the understanding as for the hand. And as the instruments of the hand either give motion or guide it, so the instruments of the mind supply either suggestions for the understanding or cautions.[1] (Book I, Aphorismus 2).

Hopefully this book can give some contribution to enter further into the fascinating interactions between hand, artefacts and mind within the social, technological and cognitive environments where we live today.

Ferdinando Arzarello
Università di Torino
Italy

References

Bacon, F. (1620). *Novum organum scientiarum*, Liber primus. Aphorismi de interpretatione naturae et regnohominis. Retrived from: http://www.thelatinlibrary.com/bacon/bacon.liber1.shtml

English standard translation of James, S., Robert L.E., and Douglas Denon Heath in *The Works* (Vol. VIII), published in Boston by Taggard and Thompson in 1863: The New Organon, or true directions concerning the interpretation of nature, Book 1. Retrieved from: http://www.constitution.org/bacon/nov_org.htm

Higgins, S., Xiao, Z., & Katsipataki, M. (2012). *Impact of digital technology on learning: A summary for the education endowment foundation*. Education Endowment Foundation. Retrieved from: http://larrycuban.files.wordpress.com/2013/12/the_impact_of_digital_technologies_on_learning_full_report_2012.pdf

Mishra, P., & Koehler, M.J., (2006). *Technological pedagogical content knowledge*: A framework for teacher knowledge. Teachers college record, 108(6), 1017–1054.

PISA in Focus n.64 (2015/07). Are there differences in how advantaged and disadvantaged students use the Internet? Retrieved from: http://www.oecd-ilibrary.org/education/are-there-differences-in-how-advantaged-and-disadvantaged-students-use-the-internet_5jlv8zq6hw43-en

[1]Nec manus nuda, nec intellectus sibi permissus, multum valet; instrumentis et auxiliis res perficitur; quibus opus est, non minus ad intellectum, quam ad manum. Atque ut instrumenta manus motum aut cient aut regunt; ita et instrumenta mentis intellectui aut suggerunt aut cavent.

Contents

Introduction: Innovative Spaces for Mathematics Education with Technology

Francesca Ferrara, Eleonora Faggiano and Antonella Montone

The idea of this book arose from research encounters occurred during a past ICTMT Conference: the International Conference on Technology in Mathematics Teaching. The ICTMT Conference, which is now moving to its 13th edition, has a strong commitment to promote technology in mathematics education for improving the quality of teaching and learning by effective use of technology. In recent years, international research in mathematics education has offered a range of theoretical perspectives that attempted to provide different and interrelated frames and viewpoints to the study of use and role of digital technologies in/for teaching and learning mathematics (e.g. Hoyles & Lagrange, 2010; Drijvers, Kieran, & Mariotti, 2010; Drijvers, Tacoma, Besamusca, Doorman, & Boon, 2013). But still, the integration of technology in the didactical practice, far from becoming a reality in the mathematics classroom, is a crucial issue of this discourse subjected to various lines of flight and critical interpretations.

Today, in particular, part of the discourse sheds some light on change and transformation implicated for the classroom practice of the mathematics teacher in the digital era (see Clark-Wilson, Robutti, & Sinclair, 2013). Other part mainly focuses on the influential affordances of software environments or devices (for example, Hegedus & Moreno-Armella, 2008; Arzarello, Ferrara, & Robutti, 2012;

The original version of this chapter was revised: For detailed information please see Erratum. The erratum to this chapter is available at https://doi.org/10.1007/978-3-319-61488-5_12

F. Ferrara (✉)
Università di Torino, Torino, Italy
e-mail: francesca.ferrara@unito.it

E. Faggiano · A. Montone
Università di Bari Aldo Moro, Bari, Italy
e-mail: eleonora.faggiano@uniba.it

A. Montone
e-mail: antonella.montone@uniba.it

E. Faggiano et al. (eds.), *Innovation and Technology Enhancing Mathematics Education*, Mathematics Education in the Digital Era 9,
DOI 10.1007/978-3-319-61488-5_1

Calder, 2015), as well as on the ways that technologies affect or even change the nature of the mathematical objects and relations (e.g. Rotman, 2008; Sinclair, de Freitas, & Ferrara, 2013). Generations of research on new technologies in mathematics education have been discussed (Sinclair, 2014; Drijvers, 2015). In this respect, are underlined the massive changes in the nature of the physical interactions with digital technologies, from the entirely alphanumeric hegemony of the keyboard to the even continuous movement with the mouse, to touch and haptic experiences, which put forward more direct action and gesture that come to replace the mouse and keyboard, and how these new forms of interaction make very different demands on the body but also on the mathematics, eventually implying new ways of sensing and making sense, a new kind of sensory politics at play in the mathematics classrooms (cf. chapter six in de Freitas & Sinclair, 2014).

More recently, researchers have started drawing attention to how the technology might offer new ways of engaging with mathematical thinking and engender new kinds of mathematical experiences for learners (e.g. Santi & Baccaglini-Frank, 2015; Calder & Campbell, 2016; Hegedus & Tall, 2016; Sinclair, Chorney, & Rodney, 2016). Student interactive engagement with mathematics, motivation and level of interest have also been part of the wide landscape (Attard & Curry, 2012; Lange & Meaney, 2013).

However, when the research lens is trained on educational research, the emphasis can shift away from practice and activity, from task design, from the role of the teacher but also, and more importantly, from the conceptual or empirical *positionings* of the researchers (see Herbel-Eisenmann, Wagner, Johnson, Suh, & Figueras, 2015) and the implications for their particular ways of speaking about research on technology. Beyond the fact that the nature of these positionings is revealed to vary immensely in the literature, Herbel-Eisenmann and colleagues underline that often, the sources from which people draw as they position each other are not explained well.

This fuzziness further complicates the relationship between positioning and storylines that are offered to readers in discourse, and entails a social meaning that depends upon the positioning of the speaker(s) as soon as this is seen as a product of the social force implicated in any communication action. Deepening the discussion, we might reconsider how the widespread use of technologies in everyday life has forged changes in the ways in which people interact and communicate beyond how they know, implying in turn a wide open range of possibilities for ways of positioning.

This challenging view is typical of current research in our field and the chapters in this book attempt to face such a sociological change drawing on the issue of innovation regarding researching about technologies and mathematical practice. It is also concerned with the reason under our choice, as Editors of the book, of talking about *spaces*. The image of space grasps here a vision of how the world looks to an individual and how the individual lives in the world. The spaces we take into account here are those where the authors of the different chapters live their specific perspective on innovation and technology at a meta-level, which is that of the particular researcher who is culturally positioning herself with respect to a certain perspective.

Therefore, the readers of the book can discover and recognize ideas and meanings of innovation as they emerge from the entanglement of the researchers with the mathematical activity, the teacher training program or practice, the student learning and engagement, or the research method that they are telling stories about. The multiple views that arise from this book have to be considered as a rich bundle of heterogeneous theoretical or empirical positionings of research, being them philosophical, instrumentalist, cognitive, technological or of other kind.

Starting a journey through the text, the reader will first encounter an opening scenery (Chap. From Acorns to Oak Trees: Charting Innovation Within Technology in Mathematics Education, by Carreira and colleagues) that recalls the ideas coming from the past ICTMT conferences, launching the delicate and subtle issue of how we have been used to speak of innovation within technology in mathematics education research, highlighting the few key innovations that have been seeded and taken root within the community of participants through the history of ICTMT.

After this scenery, the book is split into three parts that breach into spaces as explicit ways of positioning and telling stories about the teaching and learning of mathematics with technology.

The first part (*New spaces for research*) consists of three different chapters that advance fresh theoretical and methodological positionings about innovative ways of learning. Sinclair and Coles (Chap. Returning to Ordinality in Early Number Sense: Neurological, Technological and Pedagogical Considerations) propose to relate inclusive materialism and enactivism in concert with recent findings of neuro-science in order to think of new methodological possibilities for thinking of the importance of ordinality in the early learning of number and how this might be fostered by a new technology. De Freitas and colleagues (Chap. The Coordinated Movements of a Learning Assemblage: Secondary School Students Exploring Wii Graphing Technology) position from the perspective of assemblage theory to study how human bodies collaborate and assemble with technology when exploring mathematical ideas, offering the idea of learning assemblage to analyse data less in terms of tool use and more in terms of the affective force of the technology. Robotti and Baccaglini-Frank (Chap. Using Digital Environments to Address Students' Mathematical Learning Difficulties) centre their positioning on literature mainly coming from cognitive psychology, which helps address the issue of learning in relation to students with learning difficulties and to software that might promote new learning in this situation. Therefore, the context is different among the three chapters, but they share common interests in how specific positionings make different demands on the body and on mathematics.

The second body of three chapters (*New technological spaces*) contributes to the discourse with attention mainly drawn to affordances and innovative uses of new digital technologies. In this case, the positionings of the various researchers have in common their tentative dwelling upon implications and benefits of the technological environments. Through a comparative research, the instrumental positioning of Thomas and colleagues (Chap. Innovative Uses of Digital Technology in Undergraduate Mathematics) centres on the new use of digital environments in first year mathematics courses at the university, in order to tackle with possible discontinuities in the transition from secondary to tertiary education. Concerning the duo of artefacts designed by Maschietto and Soury-Lavergne (Chap. The Duo "Pascaline and e-Pascaline": An Example of Using Material and Digital Artefacts at Primary School), innovation is unfolded along two dimensions: the emergent relationships between the digital and the physical in the duo, and the possibility of integrating the digital in primary school in a way that supports teaching and learning practices. Weigand's contribution (Chap. What Is Or What Might Be the Benefit of Using Computer Algebra Systems in the Learning and Teaching of Calculus?) positions from the side of previous research on computer algebra systems and tries to deal with a new understanding and vision of the benefits of using this technological environment in the mathematics classroom, through the development of a competence model.

The third part of the book (*New spaces for teachers*) only includes two chapters that both shift the focus of our discourse specifically on teaching and take strong positionings on teaching as an activity. Kynigos (Chap. Innovations Through Institutionalized Infrastructures: The Case of Dimitris, His Students and Constructionist Mathematics) suggests to reflect on the potential for innovation made possible by connecting different kinds of innovation, and to re-think constructionism as an innovative activity that is rich in opportunities for meaning making in the era of large portal and the social web, through an example of constructionist mathematical activity by one teacher and his class using portals. Despite the widespread availability of new digital expressive and communicative possibilities, Tabach and Slutsky's positioning (Chap. Studying the Practice of High School Mathematics Teachers in a Single Computer Setting) differs from Kynigos', calling attention for the specific situation in which students do not have access to the digital but only the classroom teacher is equipped with a computer and data projector, therefore, pointing out the need for a new—for this reason, innovative—instrumental framework able to address and support teacher practice adequately in such situations.

The mosaic of the varied research that features this book is completed with the closing scenery (Chap. Digital Mazes and Spatial Reasoning: Using Colour and Movement to Explore the 4th Dimension). De Freitas affords to propose new inventive learning about spatial reasoning and spatial sense in four dimensions with digital maze technology, pointing to possible directions for future research on innovative approaches to mathematics thinking.

With this panorama in mind, we hope to leave the reader with a flavouring will for unfolding and unveiling—possibly, traversing—multiple dimensions of the spaces discussed throughout the book. In a way similar to how technology prompts interaction and how the teacher can create her own space for interaction, we hope that this book might contribute to current discussions on mathematics education with technologies offering researchers and readers spaces for communication and comparison and prompting them to create their own new spaces, rich in positionings and stories.

References

Arzarello, F., Ferrara, F., & Robutti, O. (2012). Mathematical modelling with technology: The role of dynamic representations. *Teaching Mathematics and its Applications, 31*(1), 20–30.

Attard, C., & Curry, C. (2012). Exploring the use of iPads to engage young students with mathematics. In J. Dindyal, L. Cheng, & S. Ng (Eds.), *Proceedings of the 35th Annual Conference of the Mathematics Education Research Group of Australasia* (pp. 75–82). Singapore: MERGA.

Calder, N. (2015). Apps: Appropriate, applicable and appealing? In T. Lowrie & R. Jorgensen (Eds.), *Digital games and mathematics learning: Potential, promises and pitfalls* (pp. 233–250). Dordrecht: Springer.

Calder, N., & Campbell, P. (2016). Using mathematical apps with reluctant learners. *Digital Experiences in Mathematics Education, 2*(1), 50–69.

Clark-Wilson, A., Robutti, O., & Sinclair, N. (Eds.). (2013). *The mathematics teacher in the digital era: An international perspective on technology focused professional development*. Dordrecht: Springer.

de Freitas, E., & Sinclair, N. (2014). *Mathematics and the body: Material entanglements in the classroom*. New York: Cambridge University Press.

Drijvers, P. (2015). Digital technology in mathematics education: Why it works (or doesn't). In S. J. Cho (Ed.), *Selected Regular Lectures from the 12th International Congress on Mathematical Education* (pp. 135–151). New York: Springer.

Drijvers, P., Kieran, C., & Mariotti, M. A. (2010). Integrating technology into mathematics education: Theoretical perspectives. In C. Hoyles & J.-B. Lagrange (Eds.), *Mathematics education and technology—Rethinking the terrain: The 17th ICMI Study* (pp. 89–132). New York: Springer.

Drijvers, P., Tacoma, S., Besamusca, A., Doorman, M., & Boon, P. (2013). Digital resources inviting changes in mid-adopting teachers' practices and orchestrations. *ZDM Mathematics Education, 45*(7), 987–1001.

Hegedus, S., & Moreno-Armella, L. (2008). From static to dynamic mathematics: Historical and representational perspectives. *Educational Studies in Mathematics, 68*(2), 99–111.

Hegedus, S., & Tall, D. (2016). Foundations for the future: The potential of multimodal technologies for learning mathematics. In L. D. English & D. Kirshner (Eds.), *Handbook of international research in mathematics education* (3rd ed., pp. 543–562). New York: Routledge.

Herbel-Eisenmann, B., Wagner, D., Johnson, K. R., Suh, H., & Figueras, H. (2015). Positioning in mathematics education: Revelations on an imported theory. *Educational Studies in Mathematics, 89*(2), 185–204.

Hoyles, C., & Lagrange, J.-B. (2010). *Mathematics education and technology—Rethinking the terrain: The 17th ICMI Study*. New York: Springer.

Lange, T., & Meaney, T. (2013). iPads and mathematical play: A new kind of sandpit for young children. In B. Ubuz, Ç. Haser, & M. A. Mariotti (Eds.), *Proceedings of the Eight Congress of the European Society for Research in Mathematics Education* (pp. 2138–2147). Ankara, Turkey: Middle East Technical University.

Rotman, B. (2008). *Becoming beside ourselves: The alphabet, ghosts, and distributed human beings*. Durham: Duke University Press.

Santi, G., & Baccaglini-Frank, A. (2015). Forms of generalization in students experiencing mathematical learning difficulties. *PNA, 9*(3), 217–243.

Sinclair, N. (2014). Generations of research on new technologies in mathematics education. *Teaching Mathematics and its Applications, 33*(3), 166–178.

Sinclair, N., Chorney, S., & Rodney, S. (2016). Rhythm in number: Exploring the affective, social and mathematical dimensions of using *TouchCounts*. *Mathematics Education Research Journal, 28*(1), 31–51.

Sinclair, N., de Freitas, E., & Ferrara, F. (2013). Virtual encounters: The murky and furtive world of mathematical inventiveness. *ZDM Mathematics Education, 45*(2), 239–252.

Part I
Opening Scenery

From Acorns to Oak Trees: Charting Innovation Within Technology in Mathematics Education

Susana Carreira, Alison Clark-Wilson, Eleonora Faggiano and Antonella Montone

Abstract Technology has created an expectation in all levels of education that requires us to understand how we can harness its potential for improving the depth and quality of mathematical learning. It is highly unlikely that there is a universal recipe or formula for how technology should be used that would satisfy every context or culture, but there have been recurring trends in the process of designing and implementing such innovative environments. By considering the papers included in proceedings of the past International Conferences on Technology in Mathematics Teaching (ICTMT), this chapter aims to highlight how a few key innovations have been seeded and taken root within this community. We begin by describing the ways in which innovation has been presented at ICTMT conferences with a view to exploring this from the perspectives of technology designers, researchers and teachers/lecturers from all levels of education. Given the extensive literature on this topic, it is not feasible to carry out a comprehensive survey of the complete literature base, however it is anticipated that the analysis of key ICTMT papers will be sufficient to present an informative and insightful picture and highlight some important knowledge and experience that has been elicited and disseminated.

S. Carreira
University of Algarve, and UIDEF, Faro, Portugal
e-mail: scarrei@ualg.pt

S. Carreira
Institute of Education, University of Lisbon, Lisbon, Portugal

A. Clark-Wilson
University College London, London, UK
e-mail: a.clark-wilson@ucl.ac.uk

E. Faggiano · A. Montone (✉)
Università di Bari Aldo Moro, Bari, Italy
e-mail: antonella.montone@uniba.it

E. Faggiano
e-mail: eleonora.faggiano@uniba.it

E. Faggiano et al. (eds.), *Innovation and Technology Enhancing Mathematics Education*, Mathematics Education in the Digital Era 9,
DOI 10.1007/978-3-319-61488-5_2

1 Introduction

The first Conference on Technology in Mathematics Teaching (ICTMT) took place in Birmingham UK, in 1993, under the stewardship of Professor Bert Waits of the University of Colombus, Ohio, in an attempt to develop an international conference series in the spirit of the US-based Conference for Technology in Collegiate Mathematics, which Bert and his colleague Frank Demanna had first chaired in 1988. The late 1980s and early 1990s was a period of great innovation for technology within mathematics education, which saw the introduction of the first handheld graphing calculators, the development of computer algebra languages such as Derive, Mathematica and MatLab, programming languages such as LOGO and the emergence of dynamic geometry software (DGS) such as Cabri-Géomètre and The Geometer's Sketchpad, which were greatly assisted by the design of the computer mouse.

For this first conference, contributions were sought in the form of peer-refereed papers that would be offered as presentations, shorter papers offered as workshops and thematic topics for a series of symposiums. In each case, the contributors were asked to submit their proposals according to one of the three conference themes:

1. The mathematical content of teaching and learning environments.
2. Technology as a resource for the teacher.
3. Hands-on interactions between learners and technology.

In addition, the contributors were asked to classify their contributions according to the level of educational interest (Primary teachers, Secondary teachers, Teacher-Educators and Lecturers in Higher Education) and to highlight whether their paper reported aspects of Research, Development and Teaching.

The subtlety of the decision by the original scientific committee to welcome contributions beyond traditional papers that reported empirical research was an important one—and one that has shaped the content and ethos of all of the subsequent ICTMT conferences since 1993. By welcoming papers that reported aspects of Development and Teaching, it opened the ICTMT conference to embrace innovation as participants sought to share their designs of technological tools/environments and/or educational courses, teaching materials and reports on teaching/learning outputs as 'work in progress'. Consequently, although the academic rigour, systematic evaluation and reporting of some of these early innovations was as yet undetermined, it gave the opportunity for the conference community to share new developments at an earlier stage.

It is in this context that we have chosen to examine the way in which important technological innovations within the field of mathematics education have been seeded in the form of acorns, taken root and, in some cases, grown into mighty oak trees. We did this by means of a historical review of the ICTMT proceedings (see Appendix 1) in which we foregrounded the plenary talks to identify some seminal themes and, following this, we scrutinised the peer-reviewed paper submissions[1] and workshop session topics to chart how these themes evolved over the subsequent years.

Drawing on the perspectives that the plenary speakers have proposed throughout the ICTMT conferences and in line with findings from significant texts in the field (e.g. Hoyles & Lagrange, 2009; Hoyles & Noss, 2003; Kaput, 1999; Shaffer & Kaput, 1999; Moreno-Armella, Hegedus, & Kaput, 2008; Laborde & Sträßer, 2010) we have identified the following themes that have proved to be significant with respect to the path of innovation in the field. These are:

1. The concept of a mathematical 'figure' and the new action of 'dragging'.
2. Multiple representations in mathematics (2-Dimensional and 3-Dimensional).

This chapter is structured as follows. After a short explanation of how we interpret the idea of 'innovation', we take each of these themes in turn and, drawing on key contributions to the ICTMT conference series, highlight how the ideas have evolved in the field.

In our final section, we build on these themes to describe some of the innovative technologies that have been showcased at ICTMT conferences that have particularly impacted on how we interact with, and communicate about, mathematics. For example, touch screen technologies, video, e-books, online communities and competitions—all set against a back drop of increasingly ubiquitous access.

2 Defining Innovation from the ICTMT Conference Series Perspective

Any historical review should take account of the context in which the reported phenomena occurred and, in beginning a historical review on technological innovation in mathematics, we have to cast our minds back to the typical classroom or research environment of the early 1990s. In these pre-internet days, most technology was 'stand-alone', although locally networked computer 'suites' were emerging. The data projector (and interactive whiteboard) were still in development, meaning that the whole class display of a computer screen was still a challenge, although some classrooms achieved this using television screens. The transfer of files was a physical one, involving computer discs and the computer mouse was becoming a 'standard' peripheral device for the newer computers.

As a conference series on technology in mathematics teaching, one might imagine that there is a clear and widespread understanding of what is meant by 'technology' amongst the community of conference participants. Unsurprisingly, the word technology is one of the most frequent words to appear in the written proceedings; indeed, the range of meanings for the word is vast and it includes:

[1]For one of the early conferences (ICTMT2, Edinburgh 1995) we did not have access to the complete peer-reviewed papers, only the accepted abstracts.

- Technology that is the platform through which the digital media is accessed and/or made visual—this has traditionally been called the 'hardware'. For example, the Personal Computer, Laptop computer, iPad, mobile phone, graphing calculator, and data logger.
- Technology that provides the means through which teachers and learners access some mathematics content—traditionally the software, or more recently applets and widgets. Examples include classes of software such as: dynamic graphing software (Autograph, Mouseplotter); computer algebra software (Mathematica, MathCad, Wolfram Alpha, Derive); dynamic geometry software (DGS) (Cabri-Géomètre, The Geometer's Sketchpad) and more recently software packages that combine mathematical representations and functionality (TI-Nspire, GeoGebra, TinkerPlots).
- Technology that communicates mathematical content as either a one-way (video, informative web-pages) or two-way (intelligent support) medium.
- Technology that combines some or all of the above to create portals (Census at School), 'e-learning environments' or support 'blended learning'.

These definitions seem to indicate a natural timeline as the technology often appears first, followed by classroom experimentation and then, in time, resulting in the design of more substantial courses and assessment approaches that bring the technology to more classrooms. However, this could be a slow process as, in the early days of technology use in education, researchers tended to adopt positivist approaches that used formal scientific methods that did not take full account of the diversity of teachers, students, classrooms and cultures, and variables involved. For example, by using a control group methodology and identical pre- and post-testing protocols for each group, particular effects on the learning outcomes like the role of the teacher, previous knowledge of the students or issues related to the underlying technological and pedagogical environment were often overlooked. This led quite quickly to the conclusion that many mathematical technologies changed the nature of the mathematical knowledge that was being taught—and so required different methodological approaches (Artigue, 2002). This resulted in the growth of the use of *design-based research* approaches (diSessa & Cobb, 2004) that integrate the processes of the design of the technology alongside systematic evaluation in partnership with stakeholders and, as a result, can fast-track the latter stage by adopting simultaneous design innovation.

2.1 Innovation in the Design and Evaluation of Technological Tools

The ICTMT conference series has always attracted participants whose primary interest has been in the design of technological environments for the learning and teaching of mathematics. Often, the conferences have been used as major 'design showcases' at which new features and functionalities are shared and conference participants have often been the first to experience such innovations. For example, Cabri-Géomètre (ICTMT1, Birmingham, 1993), Autograph (ICTMT6, Volos, 2003), Mathematica (ICTMT7, Bristol, 2005), Casyopée (ICTMT8, Hradec Kralové, 2007), CabriElem (ICTMT10, Portsmouth, 2011), and TouchCounts (ICTMT11, Bari, 2013).

There have always been tensions within the education community concerning the relationships between all those involved in technology design. Many technological products emerge from the academic research community but, as they prove their efficacy and develop their user base, it often becomes necessary to partner with or create commercial enterprises in order to market and distribute the resource more widely. Alternatively, academics and teachers collaborate with existing technology companies to develop and evaluate products with a commercial aim. In both scenarios, the respective aims of educational researchers, teacher educators, technology designers, teachers and students are of high importance as each partner seeks to maintain its principles and values. For example, the technology designer might seek to implement a new functionality because it is technically possible, whereas a teacher or researcher might be more concerned with how such functionality might influence or change the mathematical knowledge and its associated pedagogy.

The collaborations between the different people involved in the design and evaluation of educational technology for mathematics has led to a number of theoretical ideas that have supported an understanding of the ways in which teachers and learners begin to make sense of and use such innovations. For example, the important theoretical constructs of *instrumentation/instrumentalisation* (see Verillon & Rabardel, 1995 and, within mathematics education, Guin & Trouche, 1999), *situated abstraction* (Noss & Hoyles, 1996), *structuring features of classroom practice* (Ruthven & Hennessy, 2002) and *semiotic mediation* (Bartolini-Bussi & Mariotti, 2008).

De Freitas, Ferrara and Ferrari (Chap. The Coordinated Movements of a Learning Assemblage: Secondary School Students Exploring Wii Graphing Technology) bring in a new theoretical perspective to analyze and conceive the relation between the user and the tool: ('assemblage theory furnishes innovative ways of thinking about individual human bodies and how they come together with technology'). On the other hand, this perspective brings to light an aspect that did not appear overtly in our revision of ICTMT papers: technology and its relation to gamification paradigms and, eventually, games, including interactive digital games that make extensive use of sensorial elements. Indeed, de Freitas and colleagues say: "This research makes use of technology that is related to the game console Nintendo Wii because of the potential that it offers in terms of playing games through proprioception and kinaesthesia. The devices under consideration are the remote controllers (also called Wii Remotes, or Wiimotes) and the Balance Board of the Wii. The remote controllers are devices with which users can control and play games where real movement simulations are produced. The Wii balance board is usually used for games that depend on balance and body perception in space".

2.2 *Innovation in the Design and Evaluation of Classroom Tasks*

In reviewing the proceedings, it is not always possible to distinguish aspects of the design and evaluation of the technological tool from those concerning the design and evaluation of the associated mathematical task(s). Consequently, we interpret this aspect of innovation as that which is predominantly reported by authors who have taken technological tools that have been created by others within which to design classroom tasks that use the inherent software functionality. For example, a high school teacher proposing a range of mathematical tasks for secondary students using the Microsoft Excel programme (see the paper by Broman, in Fraunholz, 1997).

3 Some Key Innovations in Mathematics Education

3.1 *The Concept of a Mathematical 'Figure' and the New Action of 'Dragging'*

The advent of the computer graphic image that could be constructed and acted upon using the computer mouse, graphics tablet and pen, and more recently figure or stylus driven touch screens, was a major technological development that took place in the late 1980s and early 1990s. Within mathematics education, this led to the development of dynamic geometry software (DGS) (The Geometry Inventor, Cabri-Géomètre, The Geometer's Sketchpad, Cinderella, GeoGebra), a key feature of which was the functionality to use a set of geometric construction features to define 2-D (and also 3-D) 'figures', which could be transformed by selecting constituent parts of the figure and 'dragging' with the computer mouse. This was a revolutionary new way of experiencing a mathematical environment that extended the idea of a mathematical 'drawing'—a mathematical sketch that, although not necessarily accurate, incorporated mathematical properties. In her seminal plenary during the first ICTMT, Colette Laborde offered the following scenario to indicate how a 'Cabri figure' could be constructed in different ways to enable different points to be free to move on the circle (see Fig. 1).

As software developed over time, so the 'drawing' of a function such as $y = x^2$ might include key features such as set of perpendicular axes, a curved graph (the shape of a parabola) and with a minimum at (0,0). In dynamic technological environments, such drawings can become figures whereby classes of mathematical objects (geometric shapes, geometric scenarios, algebraic functions, etc.) can be constructed and interacted upon by 'dragging'.

In this section, we explore 'figures and dragging' from the perspective of the initial design and evaluation of the associated technologies before then considering the innovative design and evaluation of classroom tasks within such environments.

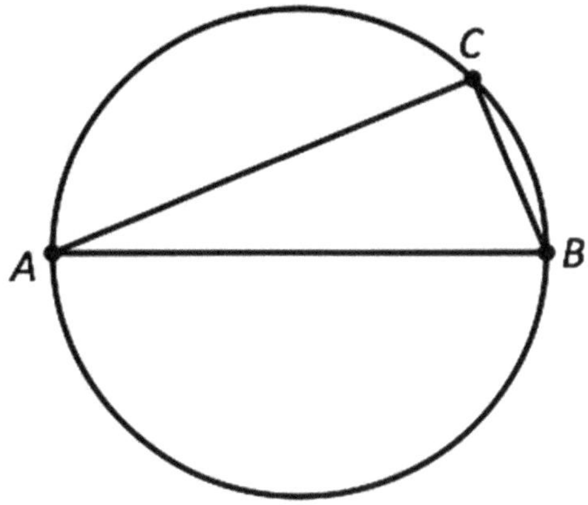

Fig. 1 The right triangle (Laborde in Jaworksi, 1993, p. 44)

3.1.1 Figures and Dragging from the Perspective of the Innovative Design/Evaluation of Tools

In her plenary address at the first ICTMT conference, Colette Laborde, who was a member of the Cabri-Géomètre design team, shared this innovative new software with the conference participants and, in doing so, offered some important definitions that were seminal in their influence of later research and practice (Laborde, in Jaworski, 1993). In the paper that accompanied her plenary, Colette made the important distinction between the idea of a 'drawing', which is 'imperfect' in that ('the lines have a width, the straight lines are not really straight') (ibid. p. 41), and the 'idealized' drawing of the mathematician, in which such imperfections are ignored through to the consideration that, in a computerised environment, 'to what extent the imperfections of a drawing are considered a noise by the users?'. Within the context of dynamic geometry, Colette then introduced the idea of a (Cabri-Géomètre) 'Figure' as a particular type of mathematical drawing which, due to its dynamic construction based on a set of (Euclidean) geometric rules, retained its mathematical properties when free objects were varied by dragging. The open nature of the construction and transformation functionality enabled a multitude of geometric figures to be constructed, from the simple case of a dynamic isosceles triangle to more complex geometric scenarios that model known and unexplored problems from traditional geometry.

In the evolutionary path of dynamic geometry tools, the dragging functionality is a major developmental milestone that was evident from the early design and, with time, has become an essential feature of most subsequent mathematical digital technologies.

Some subsidiary developmental milestones include:

- The translation of traditional geometry tools to their digital equivalences.
- A move from plane geometry to the design of 3-D geometric environments.

- The overlaying/underlying of a Cartesian plane—leading to the functionality to drag objects within multi-representational mathematical environments.
- The use of parameters and sliders to support mathematical modelling.
- The use of video/images as a context for modelling mathematics in a dynamic way.
- The use of macros (Cabri) or tools (Sketchpad, GeoGebra), which enable procedures to be captured in a process that has a strong resonance with computer programming.

The combination of dragging functionality alongside the digital representations of Euclidean geometry tools on the computer screen (ruler, compass, straight-edge and protractor) may be considered to be at the heart of the development of DGS. These capabilities have provided a new way of interacting with geometrical objects and their properties, namely by enhancing visualization, understanding, and a providing motivation for the analysis and discovery of geometrical properties. The designers enthusiastically pointed to opportunities to make the teaching and learning of mathematics more lively, relevant, fun, appealing and stimulating, particularly by giving learners 'visual hooks to hang on' (Butler, in Triandafillidis & Hatzikiriakou, 2003).

Simultaneously, from the viewpoint of teachers and researchers, there is a clear perception that DGS has great potential to emphasise conceptual understanding in geometry. For example, giving the perspective of a classroom teacher, Clark-Jeavons (in Borovcnik & Kautschlitsch, 2002a) gave an a priori synthesis of a variety of new approaches centred on conceptual understanding, not just related to the production of dynamic and interactive representations (the pointer as an extension of the hand through the mouse interface) but also profiting from complementary features. The ways in which DGS is effective in developing geometrical understanding concerns 'visual creation and interpretation', 'support in deductive proof', 'means for making and testing conjectures', 'black box activities', 'visual proof', and 'reinterpretation of the static geometry'.

The early versions of dynamic geometry environments sought to simulate and emulate the purity of Euclidean geometry and its structure, which translated into highly robust mathematical constructions being created on the computer. The robustness of those constructions, such as polygons with certain characteristics or other figures where points, lines or segments were dependent on certain primitive objects, was markedly one of the capabilities that mathematicians, researchers and mathematics educators considered to be both innovative and important. But for teachers, purism was soon balanced with the possibility of creating and using macros that allow users (for example, younger students) to perform rigorous constructions (e.g. to create an equilateral triangle) without having to go through all of the construction steps each and every time. As an example, Don Hoyle, in his account of his educational experience with school students learning about families of quadrilaterals and their classification (in Maull & Sharp, 1999), mentioned that students and also teachers found some of the 'more pure' versions of DGS more difficult to use. Therefore it favoured the availability of plugged-in features that allow pre-designed constructions as opposed to the classical Euclidean way of doing them. ('The first version of Cabri-Geometre was very purist, in that the only tools you had were, as in pure Greek Geometry, a ruler and a compass') (ibid. p. 1). Alongside this, there was an acceptance that it is not always desirable or necessary for students to create figures from scratch. Tasks are as valid when students are required to work on figures already constructed by acting upon them or changing them in some way, as a means of revealing important properties and features of those figures.

One significant paper on the evolution of DGS and its design principles was given by Elschenbroich (in Borovcnik & Kautschlitsch, 2002a) who emphasised that the main affordance of such environments was the opportunity to investigate invariants, functional dependencies and loci. He asserted that the drag mode and the revealing of loci are two important functionalities that extended those of (digital) construction using compass, straightedge, ruler and protractor. The resulting digital figures are classes of drawings holding the same properties and relationships and Elschenbroich argued that the purpose of teaching and learning geometry should be directed to investigate what changes and what remains invariant. Moreover, as measures can be made dynamically, it became possible to calculate with such dynamic measures and even to create new constructions using those measures. Therefore, the new dragging action was extended to coordinates, measurements and equations. And the figures could also go beyond the user's constructions as it became possible to produce them under the form of loci. Dragging a point could be interpreted as leaving a visible trace and thus generating a new object—a locus. Such loci remained available as new objects that could lead to new mathematical investigations. The meaning of dragging thus became much broader than it was before.

Another storyline concerns the evolution of technological tools from plane geometry to geometry in three dimensions, which became visible at the ICTMT conferences through the development of 3-D DGS. This innovation transported the dragging functionality that had been developed for 2-D DGS and made it possible for points, lines, vectors and surfaces to be manipulated in 3-D space.

Alongside this, the teaching and learning of coordinate geometry, non-Euclidean geometry and other versions of algebraic approaches to geometry gained new perspectives by appropriating the dynamic nature of the changing of drawings in a figure and the corresponding change in the equations and/or parameters involved in coordinates and Cartesian equations. For instance, Adrian Oldknow (in Borovcnik & Kautschlitsch, 2002a) commented on the advances and developments of DGS by stressing the link between geometric and algebraic representations: ('The current versions of Cabri and GSP both provide the means to use the results of measurements and calculations based on them to define the position of points in a Cartesian coordinate system. Thus they can be used as algebraic tools where a graph of a function can be created as the locus of a point whose *y*-coordinate is a given function of its *x*-coordinate') (p. 84).

One prominent innovation concerns the combination of and connections between geometric objects and algebraic objects/symbols in ways that capitalise on the dynamic properties of many of these software products alongside the nature of the underlying mathematical constructions. Roanes-Lorenzo (in Borovcnik & Kautschlitsch, 2002a), in collaboration with other Spanish researchers, described the creation of a combined package, called Lugares, that made it possible to link dynamic geometry and algebra. The author claimed that DGS and CAS packages had attained high levels of development but they had evolved independently. The aim was therefore to develop a software package that provided the equations of any drawn configurations—in particular those of parameter-dependent geometrical constructions. ('In Lugares, numerical approaches to dynamic geometry can be complemented with the symbolic capabilities of CAS, letting a step forward to draw loci and find their equations') (ibid, p. 361).

Indeed, this combination of geometry and algebraic equations in coordinate systems was promptly felt as an important innovation for the work with figures that depended on the variation of parameters. For example, innovative ways of dealing with conics (see Broman, in Borovcnik & Kautschlitsch, 2002a) or investigations on functions and their graphs in relation to the change of areas or distances began to emerge and be discussed as legitimate mathematical tasks, sometimes in tune with a problem-solving or a modelling approach to mathematics (see Lopez-Real in Olivero & Sutherland, 2005 and Miller & Ehmann, in Triandafillidis & Hatzikiriakou, 2003).

We note that by the mid 2000s, the community's knowledge and understanding of dragging (by then an essential affordance of DGS) was consolidating, and the action of dragging was becoming more and more intrinsic and even 'natural' within DGS environments. This in turn was generating ever more specific and more diverse results in terms of the kinds of representations offered and combined within mathematical tasks.

Although the development of software that enabled the progression from 2-D to 3-D contexts is an important benchmark in the process of innovation, the ability to generate loci and work with them in concrete ways, namely by dragging them, should be acknowledged as a significant step forward. The example given by Miller and Ehmann, in which the locus of the intersection points of the altitudes of triangles with constant height is described, is illustrative of this and, as the authors argue, it ('puts across a first impression of the transition from an elementary geometric problem to a question in calculus') (Triandafillidis & Hatzikiriakou, 2003, p. 315).

3.1.2 Figures and Dragging from the Perspective of the Innovative Design and Evaluation of Classroom Tasks

The ICTMT proceedings tell an interesting story with respect to the development of innovative classroom tasks that utilised the ideas of dynamic geometric figures that could be constructed and interacted upon by dragging. During ICTMT1, which featured the seminal plenary by Colette Laborde, there was only one other paper (Little, in Jaworski, 1993) that made reference to how dynamic geometry software (Cabri-Géomètre) might influence the future design of classroom tasks. Little makes specific reference to the importance of the distinction between a figure and a drawing when clarifying geometric concepts. Little writes, ('In cabri, we construct a 'figure', then by dragging vertices we see different 'drawings' which exemplify the figure. This distinction can be used to illustrate and clarify the distinction between, for example, 'square' and 'squareness''). In his paper, Little suggests such tasks might become legitimate in a technology enhanced school geometry curriculum in England.

The strong attention given by researchers, teachers and lecturers to the dragging action was reflected not only in the exploratory, inquiring and investigative nature of the tasks supported by DGS but also within the theorizing of different modes of dragging and of the role of dragging in mathematical activity involving the construction and manipulation of figures on the screen.

In her paper presented at ICTMT4, Federica Olivero (in Maull & Sharp, 1999) reported outcomes of a study with 15-year-old students in which a theoretical model of 'dragging modalities' was proposed. The study was described as follows: ('Pupils were requested to produce conjectures in open geometric situations, to validate and, finally, to prove them. These activities took place within the microworld Cabri-Géomètre. (…) We found out that different dragging modalities are crucial for producing a shift from conjecturing to proving: these modalities can be analysed as the perceptive counterpart of the cognitive processes students use'). (ibid. p. 567).

This particular study showed that students' use of dragging in Cabri changes with respect to the control that the students have of the situation. It also showed that dragging acts as a mediator of students' activity on proof. The essential argument is that dragging supports the production of conjectures, given that exploring figures by moving and manipulating them, allows the users to discover invariant properties of those figures. Besides, the possibility of dragging has another major function, that of generating useful feedback in the phase of discovering properties and furthermore supporting the finding of a proof as an 'explanation' of the property or the conjecture made.

This mediating role of the tool and especially the dragging action has continued to be investigated and still remains important in the research about the mathematical thinking afforded by DGS in tasks that clearly highlight the dragging function in deduction, explanation and proof.

It is worth pointing out that the views of teachers and researchers did not always seem to coincide with regard to evaluating the affordance of dragging. While teachers tended to look at the dragging as handy to help students overcoming difficulties in concept acquisition, researchers were theorising the dragging action and conceptualising the subtleties of this feature in students' activity from exploration into proof. For example, in the work presented by Kordaki, Balomenou, and Pintelas (in Triandafillidis & Hatzikiriakou, 2003) students were asked to construct several triangles and then to transform them into other equivalent triangles. The difficulty of the concept of equivalent triangles (conservation of the area) was the motivation for the study. The authors claimed the innovative character of this DGS approach to the conservation of the area. The main result of the teaching approach was that students ('viewed the concept of conservation of the area of a triangle as an alteration of its position on the computer screen as well as an alteration of its figure') (ibid. p. 181) and ('using the drag mode… students had the opportunity to observe a large number of equivalent triangles'), ('thereby forming a dynamic view of this concept') (ibid. p. 181). Thus the emphasis on the dragging action was related to the possibility of observing a large number of cases. In short, the power of dragging was absorbed by many teachers, in the tasks they produced for their students, as opportunities to visualise and to make sense of a concept or property, which would not be feasible to realise on paper.

3.2 The Concept of Multiple Representations

Since the late 1980's a growing appreciation of the importance of multiple representations has been considered an essential component in the process of learning mathematics. The learning or doing of mathematics implies not only manipulating mathematical symbols, but the interpretation and coordination of mathematical relationships and situations, using specialised language, symbols, images and graphs. It also involves the clarification of problems, deduction of consequences and development of appropriate tools (National Research Council, 1989).

Moreover, in the USA, the National Council of Teachers of Mathematics (NCTM, 1989) published 'Curriculum and Evaluation Standards for School Mathematics', in which multiple representations are considered as one of the fundamental aspects of the curriculum that should be emphasised during the teaching and learning of mathematics. According to the document's authors, students who have flexible tools for solving problems are able to interpret the same problem or the same mathematical concepts through its multiple representations.

Other researchers at this stage and thereafter in the 1990's were devoting much attention to the role of representations and to the importance of translations between multiple representations in students' learning and understanding (of numbers, algebra, functions, etc.) and also to the multi-representational capacity of many developing software (e.g. Fey, 1989; Schwarz, Dreyfus & Bruckheimer, 1990; Borba & Confrey, 1996).

Some of the contributions at ICTMT corroborate this trend. Gomes Ferreira (in Fraunholz, 1997) suggests that the incorporation of multiple representations in

x	2x-8
-2	-12
-1	-10
0	-8
1	-6
2	-4
3	-2
4	0
5	2
6	4
7	6
8	8
9	10

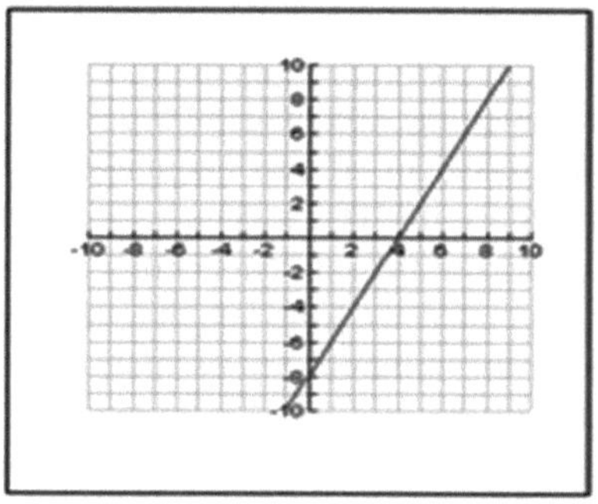

f(x) = 2x-8

Fig. 2 Multiple representations of the linear function, f(x) = 2x−8 (Adu-Gyamfi, 1993, p. 11)

mathematics teaching can empower and help students to develop their understanding of mathematical relationships and concepts and improve their perceptions of mathematics by emphasising the linkages between graphical, tabular, symbolic and verbal descriptions of mathematical relationships and mathematical problem situations.

This view is concurred by Ozgun-Koca, who says that multiple representations can be defined as external mathematical embodiments of ideas and concepts to provide the same information in more than one form (Ozgun-Koca, 1998). For example, a linear function can be viewed as a set of ordered pairs, a correspondence in a table or a mapping, a graph, or an algebraic expression. Examples of external multiple representations include verbal representations (written words), graphical representations (Cartesian graphs), algebraic or symbolic representations (equations expressing the relationship between two or more quantities), pictorial representations (diagrams or drawings) and tabular representations (table of values) among a host of others (see Fig. 2).

Innovation in technology design has led to the development of digital environments that enable multiple linked representations to be acted upon such that the relationships between mathematical representations can be made more visible. Depending upon the design of the associated mathematical tasks, students and teachers can focus their activity on establishing and justifying the underlying mathematical links and connections.

3.2.1 Multiple Representation from the Perspective of the Innovative Design/Evaluation of Tools

The analysis of the ICTMT proceedings shows that one of the most influential technological tools to use multiple representational functionalities in mathematics has been Texas Instruments' TI-92 graphing calculator. Since the early 1980s developments and applications through different experiences in school mathematics education have been reported that exploit its multiple representations. Aspetsberger (in Fraunholz, 1997) presented the outcomes of classroom work with students aged 16, who were mainly interested in art and languages and not in natural sciences. The author's goals were to use the TI-92 to make traditional mathematical content more visual and accessible for the students. For example, referring to teaching calculus and problems associated with the tangent, the author says that ('especially for optimization problems the different representation modes of the TI-92 (table, graph, expression) were very helpful for illustration. The students learned how to detect minima and maxima in tables, graphs and to verify them by means of calculus. For curve analysis the permanent availability of graphs was very illustrative') (p. 2). And, referring to the experiences, she also says ('One of the main advantages of the TI-92 are [sic] the different forms of representation (tables, graphs, expressions) which are always available on the TI-92 and can lead to a better understanding of mathematical concepts. The students have the possibility to choose a representation form they like most e.g. for solving problems, for illustration or to get an overview in a certain situation. It is remarkable, that most students choose tables or graphs to solve problems, if the method is free. Only very few students use expressions for solving problems or for illustration. The abstractness of expressions is a major handicap in traditional math courses when introducing new mathematical concepts. So the availability of different representation forms helps to differentiate and individualise the process of math teaching') (p. 12). This teacher's conclusions, supported by the evidence she elicited from her own classroom experiences are echoed by many other teachers and lecturers who have presented outcomes of their own classroom work in ICTMT presentations and workshops over the years.

Another relevant paper by Duncan (in Bardini et al. 2009) describes the design of a more recent technological tool, which also emphasises the role of multiple representations. In his paper, Duncan presents the results of a classroom-based study that evaluated the Texas Instruments TI-Nspire, which evolved from the previous TI-92 technology. Duncan, referring to the work of Richard Skemp (1978), poses the problem of transition from an *instrumental understanding*, characterised by 'rules without reasons', to a *relational understanding*, in which students understand what they are doing and why. In particular, he describes a study done in schools in Scotland during 2008/09 and reports the views of the teachers involved in this study. The use of multiple representations within TI-Nspire and the evidence of students' relational understanding associated with mathematical concepts is detected through the teachers' responses to his research questions.

The teachers were asked in a direct way whether they perceived that the use of multiple representations facilitated by their task designs within TI-Nspire had enhanced their students' relational understanding of the mathematics involved in their lessons. The majority of teachers responded positively to this question and examples of their justification were:

- ‘Pupils making connections/links between topics or single concept from different perspectives’;
- ‘Pupils explaining topic/lesson to others verbally—[their] discussion’; and
- ‘Pupils asking/answering questions—wanting to know why’.

The evolution of technological tools has also resulted in environments that encourage the development of visualization skills in 3-D. It has always been challenging for teachers to promote the development of 3-D concepts in a traditional classroom environment using the standard chalkboard to represent 3-D objects by means of a 2-D sketch. It is generally admitted that learning 3-D geometry is strongly related to wider spatial and visual ability (Dreyfus, 1991). This has led to the creation of software that aims to develop abilities and processes in students that are closely associated with a mental scheme representing spatial information.

It follows that the design of some important software followed some major fields of educational theory. On the one hand, the constructivist perspective about learning, which argues that learning is personally constructed and is achieved by designing and making artefacts that are personally meaningful (Kafai & Resnick, 1996); on the other hand, the semiotic perspective that views mathematics as a meaning-making endeavour and argues that any single sign (e.g. icon, diagram, symbol) is an incomplete representation of the object or concept, and thus multiple representations of knowledge should be encouraged during learning (Yeh & Nason, 2004).

Several researchers offer definitions of visual and spatial ability. According to both Tartre (1990) and Linn and Petersen (1985), spatial ability is defined as the mental skills concerned with interpreting relationships visually, understanding, manipulating and reorganizing, and also as the process of representing, transforming, generating non-linguistic information.

Gutiérrez (1996) considers that the development of 3-D dynamic geometry software necessitates the following core visual abilities to be taken into account: (1) *perceptual constancy*, (the capability to recognise the independence among some proprieties of an object and some characteristics such as size, colour, texture, position, different orientation); (2) *mental rotation*, (the ability to visualise a configuration in movement); (3) *perception of spatial position and spatial relationship*, (the capability to relate objects, pictures or mental images to oneself or to each other); (4) *visual discrimination*, (the ability to compare several objects, pictures, mental images to identify similarities and differences among them). Finally the access to a multitude of representations supports students to create ‘correct’ mental and spatial images.

These theoretical perspectives (and the associated elements of visualization) informed the design of an early 3-D geometry software environment, *3-DMath*. (‘The idea of *3-DMath* is to develop a dynamic three dimensional geometry microworld, which enables (i) students to construct, observe and manipulate geometrical figures in 3-D space, (ii) students to focus on modeling geometric situations, and (iii) teachers to help students construct their understanding of stereometry’). (Christou, Pittalis, Mousoulides, & Jones, in Olivero & Sutherland, 2005, p. 69).

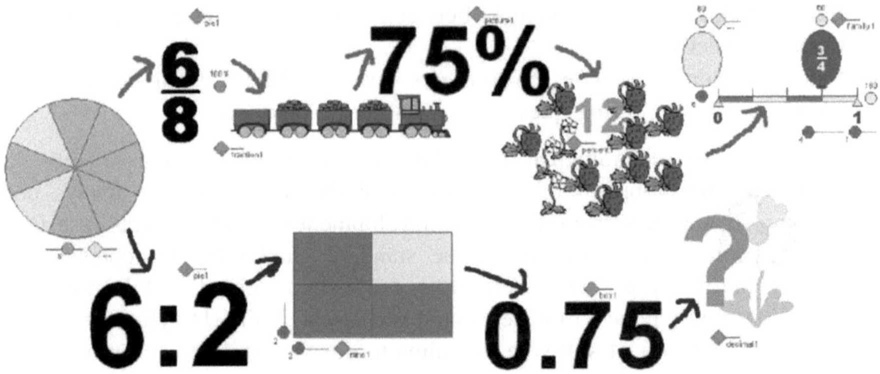

Fig. 3 Two chains of dependencies of fraction objects staring with a pie with the value of 6/8. (Lehotska and Kalas in, Olivero & Sutherland, 2005, p. 110)

The potential of multiple representations has also been exploited within the design of arithmetic environments. For example, the software Logotron Visual Fractions (LVF) was presented at ICTMT7 by Lehotska and Kalas (in Olivero & Sutherland, 2005). LVF gives students the opportunity to become aware of the meaning of the different semiotic registers and provides new learning opportunities for discovering and exploring fractions and fractional relations through the dynamic dependencies and interactions of the different representations of fractions. This complex tool provides ten different visual interactive representations of fractions and the relations between them: pie, box, decimal fraction, percentage, ratio, picture, family, number line, fraction and balloon and fraction objects can be connected together to create dependencies (see Fig. 3).

In this way it is possible that some objects represent values of some other objects. These dynamic dependencies and interactions give students and teachers the opportunity to observe what happens if we change the value of some object denoted by one of the representations and find the reasons why a fraction represented in one way cannot be expressed in another type of representation. For example, fractions represented by a picture 'story' such as a train with three of its four carriages filled with coal has a fixed denominator, which is why it cannot represent any value of another fraction. The relations between objects can also be more complex—more objects can depend upon the same object.

The importance of the visualization of different representations was also emphasised by Butler (in Triandafillidis & Hatzikiriakou, 2003) in his plenary during ICTMT6. He highlighted how the exploration and use of dynamic 3-D objects (lines, vectors, planes, etc.) within software such as Autograph v.3. can help students make sense of 3-D mathematical situations where they are required to solve problems involving the intersections of planes, and the shortest distances between points. Furthermore, he underlines that for many teachers, although software offers freedom from the more limited chalkboard, it presents a serious challenge to their well-established and already effective teaching styles.

A key feature of all the tools described so far is that they offer the opportunity to improve the construction of mathematical meanings through the use of multiple representations—an innovative and unique affordance of such technology-mediated environments.

3.2.2 Multiple Representations from the Perspective of the Innovative Design/Evaluation of Classroom Tasks and Lesson

The value of working with multiple representations has been a consistent theme in the literature for more than two decades. However, how to integrate the use of multiple representational technology into classroom tasks and how to reorganise classroom work so as to make the most of the potential of technology to enhance learning have been recurring questions for researchers and teachers/lecturers alike.

The evidence of the ICTMT proceedings highlights the dichotomy in responding to these questions. On the one hand, several researchers underline that the complexity of mathematical structures and the multiplicity of its representations within dynamic technological environments make the mathematics difficult to learn and to understand. On the other hand, researchers have shown that the students who are more successful in mathematics are those who have been exposed to multiple representations of ideas and mathematical principles. In particular, the coexistence of the three basic components, algebra, geometry and number, in situations of teaching and learning of mathematics seems to underpin such success.

The innovation offered by the multiple representations, namely the simultaneous display of different representations of the same mathematical concept (algebraic representation, geometric and numerical) opens up the possibility of restructuring the lesson, by redefining objectives and tasks. However, this presents new challenges for teachers as they begin to create mathematical tasks that use the affordances of such software environments to promote productive activity for their students. An important contribution on this theme is given by Pierce and Stacey (in Bardini et al., 2009). They describe their research that ('reports on the use of 'lesson study' to research principles for the design of a lesson aiming to use a pedagogical opportunity at the task level: the use of multiple representations') (p. 1). They identify four key principles: ('focus on the main goal for that lesson (despite the possibilities offered by having many representations available); identify different purposes for using different representations to maintain engagement; establish naming protocols for variables that are treated differently by-hand and within a machine; and reduce any sources of cognitive load that are not essential') (p. 1). Furthermore the authors have elaborated the pedagogical opportunities map, which summarises the different levels (the task, the classroom, the subject of mathematics) afforded by technologies (see Fig. 4).

Pierce and Stacey underline the importance of developing lessons to help teachers and students to gain advantage from access to multiple representations. Their study highlights the importance of technologies to allow students to explore problems in the new way using multiple representations. They suggest the need to

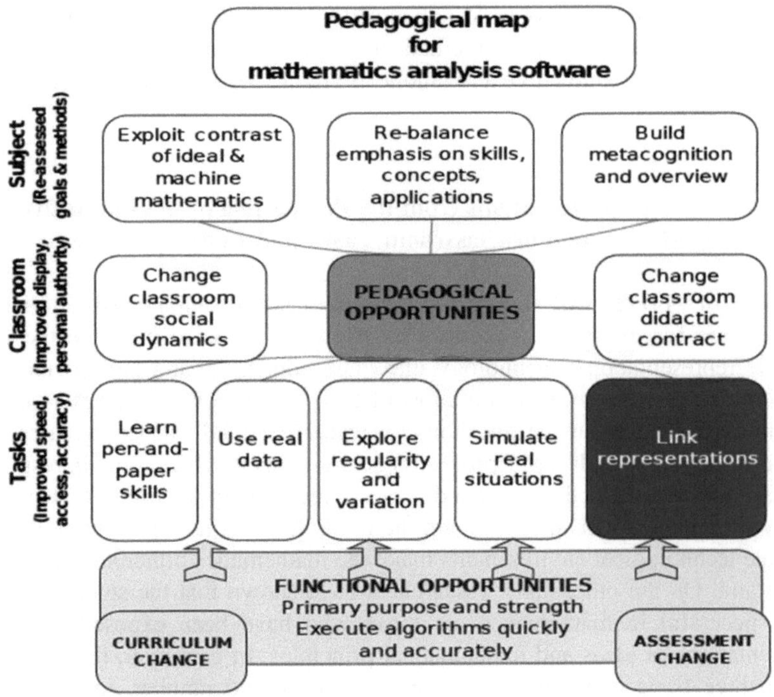

Fig. 4 Pedagogical opportunities' map, emphasising linking representations at the level of classroom task design. (Pierce and Stacey, in Bardini et al., 2009)

focus on students' thinking on a particular area of mathematics to allow teachers to restrict the strategies supported and to plan to reduce distractions due to the technology: by establishing naming protocols and minimizing the amount of extraneous information students must deal with.

4 Ways of Interacting with Mathematics Through Technology

Since the first ICTMT in 1993, the ways we think about interacting with mathematical objects have evolved in line with the development of new technologies.

The word 'interact' and its derivations is prevalent within all of the ICTMT proceedings—often appearing more than 100 times in each volume—but its use is mainly as an adjective to describe all manner of innovative technological environments (see Nodelman, in Bardini et al., 2009). There are only a few examples where authors have probed the nature of the interaction in more depth by posing questions such as: Is it an interaction with something? Is the interaction facilitated by something? Who is doing the interaction and with what?

Lapp (in Fruanholz, 1997), in his paper on students' perception of the authority provided by the technology as they interacted with TI-82 handheld calculators, offers a model of their interactions for aspects of their technology use. He also considers the importance of the role of the teacher's interactions—underlining the teacher's role in supporting the tendency for students to believe the calculator's model (or not). Lapp reported that, when faced with an apparent contradiction, the students would often refer back to the teacher's previous statements that were made during the lesson.

In her plenary talk at ICTMT7, Mariotti introduced the important theoretical point of view of *semiotic mediation* drawing on the work of Vygotsky to say ('it is possible to interpret the role of artefacts and their functioning as sources of meaning, from the educational point of view') (see Mariotti, in Olivero & Sutherland, 2005) (p. 5). She recognises Papert (1980) as the first person to acknowledge that interacting with a computer offers many different opportunities for meaningful activities and involves ways of thinking which are recognisable as typical of mathematics. However, the process of construction of mathematical meanings is not directly and simply related to the interaction with the technology. Mariotti's work contributes to the need to increase investigation on activities in computer environments in order to study the effects that such activities may have on the mathematical classroom as a whole. One of the main ideas in her paper is that the mathematical meaning related to technology becomes accessible to the learner by its use, but the construction of meanings is fostered by the guidance of the teacher. The mediating role of the teacher is essential for the interaction between the learners and the technology to be effective in terms of learning gain.

Following Texas Instruments' launch of the TI-92 handheld calculator in 1996, which incorporated a Computer Algebra System (CAS) and an interactive geometry package (based on Cabri-Géomètre), many ideas about the classroom use of CAS and DGS were seeded and began to take root as the ICTMT conferences were populated with papers and presentations on these themes. Teachers and researchers underlined how interactions with the different representations, which are always available on the TI-92 and its successor, the TI-Nspire, exploits the affordances of CAS and DGS, and can lead to a better understanding of mathematical concepts. For example, Duncan (in Joubert, Clark-Wilson, & McCabe, 2011) attributed the positive effect of the use of multiple representations on the development of students' relational understanding in large part to the interaction with TI-Nspire. The interaction with the handled technology affected students' engagement, positive attitudes and perseverance in the activities, thus enhancing a genuine, deep understanding, characterised by students knowing both what to do and why. This resonates with the work of Weigand (Chap. What Is or What Might Be the Benefit of Using Computer Algebra Systems in the Learning and Teaching of Calculus?), which energizes the many studies and research involving the multiple representations of mathematical concepts to show in a powerful way how the representational capacity is constantly increased and intensified in CAS environments like GeoGebra or the later graphing calculator. But Weigand's contribution brings another side of innovation: the awareness that there is a new competence to be developed in order to take advantage of CAS in the learning of sequences, functions, or equations. And again this will mean to rethink the work with symbolic, numerical and graphical representations so as to determine the best way that a student can be supported to attain a better understanding and higher competencies in working with CAS.

More recently students are becoming increasingly familiar with touchscreen devices (such as interactive whiteboards, tablets, smartphones, etc.) as they become more and more available within and outside the classrooms. In addition, multi-touch technologies are both impacting upon and providing challenges for mathematics education as they provide a multi-modality of forms of interaction and communication that can be enriched by new ways of manipulation. Some innovative experiences conducted with the use of the new touch technologies have been presented at the more recent ICTMTs and portend to take root in the coming years.

As an example, we refer to the work of Arzarello and colleagues (see their paper in Faggiano & Montone, 2013), in which they analysed and identified modes of touch screen use during the process of solving problems using a dynamic geometry software. Based on research by Yook (2009) and Park, Lee, & Kim (2011), and distinguishing between finger action from the user and motion feedback from the interface, they observed singularities in the way students perform rotation (using one or more than one finger) and a different way of dragging (called the 'dragging to' approach). Focusing on motion feedback as a powerful strategy to improve interaction, discovering and thinking in mathematics education, Arzarello and colleagues highlighted the new challenges of interaction and learning processes in the key transition from click to touchscreen interactions.

Jackiw (in Faggiano & Montone, 2013) imagined the glass tablet screen ('as some sort of conceptual border between the Platonic realm of geometric abstraction (on the computer's side of the glass) and the tactile empire of sense experience (on the user's side)') (p. 149). He questioned how multi-touch ideas could shape mathematical ideas and, alternatively, how multi-touch approaches could impact upon learners' interactions. Illustrating '*Sketchpad Explorer*', he highlighted that the gesture space of a multitouch device establishes a full semiotic system, in which the conditions for mathematical structure or meaning become possible. However such innovative technologies require more work to understand how to re-craft rich software interfaces toward multi-user asynchronous uses. The appeal of this paper goes beyond the designers' points of view: multi-touch enables uniquely-embodied interactions with multivariate mathematics and this offers opportunities ('to both extend and rethink existing research on embodiment in Dynamic Geometry in both individual and social formations, as well as a boldly-literal new meaning for digital mathematics') (p. 154).

At ICTMT11—and in her keynote at ICTMT12—Sinclair (see her papers in Faggiano & Montone, 2013 and in Amado & Carreira, 2015) described *TouchCounts*, an application designed for the iPad with the aim to assist young children (ages 3–8) in developing an understanding of the one-to-one relationship between their fingers and numbers. As the multimodal touchscreen interface provides direct mediation through fingers and gestures, the study shows how the application seems to facilitate the establishment of number practices and the development of number sense through both the individual and collaborative finger-based interactions. The complex analysis of the way in which young children become fluent with cardinal aspects of number while using *TouchCounts* has been developed using a new materialist theoretical lens, according to which the tool and the user mutually constitute each other through interaction (de Freitas & Sinclair, 2013). Sinclair's research, hence, represents an interesting example of another acorn which might germinate in the coming years.

The impact of these new ways of interacting and levels of connectivity are yet to be evaluated. Moreover, since 1993, the opportunities and means for the sharing of resources and ideas, and of collaborating through technological devices are maturing at great pace. Many papers, especially throughout the second half of the conference editions, deal with different aspects of the opportunities afforded by the almost ubiquitous access to information, learning materials, experts' guides, and self-assessment tools. Online courses are extremely attractive in the flexibility that they offer to learners and in recent years computer-aided learning has become a recurring addition to mathematics education. Many Virtual Learning Environments (VLEs) have been designed, implemented and experimented upon especially at the high-school and university level. High expectations seem to be placed on some of the main features of these interactive resources, such as the opportunity for students to practise and interact with mathematics anytime and anywhere and quickly receive automatic intelligent feedback on their work, or the possibility for teachers to track and review the students' progress and difficulties (see Bokhove, in Milkova, 2007).

More recently, in today's rapidly changing educational landscape, various content management systems have also been used for teachers' professional development. They seem to take advantage, in particular, of the communication functionalities which allow students and teachers to share experiences, exchange ideas and interact with other colleagues, creating a supportive and collaborative working environment (see for instance the papers presented by the MEI team members at ICTMT10 and ICTMT11 in Joubert, Clark-Wilson, & McCabe, 2011 and Faggiano & Montone, 2013, respectively).

To conclude this overview on how the many new ways of interacting with mathematics have impacted upon technology and innovation in mathematics education, a special mention should be given to the research projects that explore the affordances of e-books, which offer new kinds of flexibility, participation, and personalization. It is assumed that, as the traditional textbook will rapidly evolve from print to digital formats, the ways in which teachers and students will interact with such textbooks will also develop. In particular, the Museum Image Model of non-ordered multi-modal digital textbook as presented by Michal Yerushalmy in her keynote at ICTMT11 (see her paper in Faggiano & Montone, 2013), which constitutes an interesting example of how technology could exploit research results so as to foster the development of innovative teaching and learning tools. In her view, an e-book presents opportunities for students to focus on a concept and practice related skills, making the objectives of the learning apparent, while the interaction can be guided by the tasks, by the tools, by the feedback of the interactive diagrams, and by the problems and exercises. Yerushalmy also points out that, although some teachers call for greater participation in choosing and authoring textbooks, it is unclear whether and how teachers and schools could assume an important role in designing and developing curriculum materials, and how this would change the way they use textbooks in a sustained way.

A current European Union funded project, MC-Squared is also grappling with some of the same issues. During the most recent ICTMT, members of this project team outlined its aims: to design and develop a new genre of author-able creative e-books, called c-books, which consist of pages with carefully designed interactive elements (widgets) (see the papers by Kynigos and Kalogeria and Bokhove et al. in Amado & Carreira, 2015). Several *Communities of Interest* (Fischer, 2001) in the different participating countries, who have diverse creative profiles, fuel the collaborative design and development of the c-books within a socio-technical environment leading to the production of creative outputs. According to the early results of the project, the c-books have the potential to foster students' creative mathematical thinking and can also function as a catalyst for teacher professional development. In particular, by engaging in design activities, teachers can develop a better understanding of the relationship between technology, pedagogy and the content being taught. This is an emerging research field which may prove to be another acorn that might take root and thrive to become an oak tree.

5 Conclusions

At ICTMT conferences it is not unusual to see teachers and lecturers presenting 'innovations' in their own classroom teaching that do not appear to be so innovative to the older or more experienced members of the conference community. This does suggest that despite the existence of a large body of research on technology use in mathematics education, practitioners are more likely to begin by experimenting in their own classrooms, which in turn stimulates their interest in the existing community and its research findings. The chapter proposed by Kynigos (Chap. Innovations Through Institutionalized Infrastructures: The Case of Dimitris, His Students and Constructionist Mathematics), for example, highlights the renewed importance of teachers as co-designers of digital tools and task creators, whether challenging the capacity and features of the digital media that are meant to fit the characteristics of his/her classroom, whether exploring more freely and boldly tasks in which students can use software tools to develop diverse approaches to mathematical questions.

Maschietto and Soury-Lavergne (Chap. The Duo "Pascaline and e-Pascaline": An Example of Using Material and Digital Artefacts at Primary School) also pose a question about the way in which digital tools seek to emulate properties and rules that define mathematical objects; this therefore means bearing in mind the importance of the feedback that digital artifacts must offer to the user. Moreover, in the case of their research, innovation is about connecting physical and digital manipulatives, and in particular exploring complementarities. The study highlights the different educational settings that teachers are able to imagine and productively organize around the use of digital tools in their classes, which reveals how innovation is essentially shaped by the teacher's action with their students.

Indeed, by participating in the ICTMT conferences, many teachers and lecturers make the first step towards becoming researchers as they engage with other participants and learn about other projects and technologies that relate to their own interests. Hence, over time it is possible to see what appears to be the same presentation—albeit from a different teacher/lecturer that uses a variety of technologies—that exemplify the presenter's personal excitement in discovering the power of dragging objects in dynamic software or the affordances of multiple representational software. What is important is that, by combining and contrasting experiences, the community works to advance both knowledge and practice in an inclusive way that takes account of different starting points.

One of the most significant technological innovations in the field, the *drag mode*, has now become an inherent feature in the design of most mathematical technologies that offer multi-representational environments. Initially dragging was carried out by using the cursor keys on a computer keyboard, later using the computer mouse and now facilitated by touch screen technology; this functionality to manipulate on-screen mathematical objects has been a 'game-changer' for mathematics education research and researchers. However, repeated reports of technology use in mathematics education around the world suggest that not much has changed with respect to classroom uses of such technologies. There is still much work to be done to enable prospective and practising teachers and lecturers to experience accessible classroom tasks that enable students to interact with mathematics through the act of dragging tangible mathematical objects. Hence, each teacher and lecturer needs to become an innovator, feel the excitement of seeing his or her own students engage in purposeful, rich mathematical activity in new ways and, in turn re-evaluate their teaching approaches.

As mentioned in the introduction to this chapter, traversing through the ICTMT conference series, it is possible to identify clear tracks of innovation in the design and pedagogical uses of technological tools for the teaching and learning of mathematics at different educational levels. The characteristics that marked the evolution of educational technology stem from how deeply they have transformed the ways in which the ideas, objects and mathematical concepts are approached. The drag mode associated with the dynamic characteristics of the software is becoming the norm. The multi-representational nature of the signs and images that can be manipulated, observed, modified and connected on the screen is yet another trend that has developed as a highly relevant aspect in the teaching of mathematics. Both trends together have led to the merge of software packages. The use of increasingly portable technology has also consolidated and increased the importance given to student's independent and collaborative work in an atmosphere of problem solving, mathematical modelling and investigations, in which a key role of the teacher is guiding the student's activity with the technology. Alongside, it is noticeable that an emphasis on meaning-making and higher order thinking based on the development of mathematical concepts and its multiple representations has followed most of the technological innovations in mathematics education. In this regard, Sophocleous and Pitta-Pantazi (in Amado & Carreira, 2015) present and discuss different modalities of using the opportunities offered by technology for promoting higher order thinking in mathematics. Based on research results they claim that the most successful technological environment in improving students' mathematical learning was the inquiry based technological environment.

We anticipated that our review of the ICTMT proceedings might lead us to be able to comment on innovation from the perspective of the design and evaluation of courses and assessment processes on a larger scale for participants that include school-age students, university-age students and both pre-service and in-service teachers. The ICTMT conferences have always included designers of curriculum and assessment processes who have worked in partnership with the community to develop innovation in these areas. However, whilst carrying out our review, it was noticeable how few contributions addressed the scaling of technology mediated teaching approaches. Whilst some early contributions made reference to courses in design (see Oldknow, in Jaworski, 1993), there are far fewer that report evaluations of the outcomes of courses that have been established over time. Evidence from the more recent ICTMT proceedings suggest that researchers' attentions are moving towards the design and evaluation of:

- courses for upper secondary mathematics (Brockmann-Behnsen, in Joubert, Clark-Wilson, & McCabe, 2011; Weigand, in Faggiano & Montone, 2013)
- courses in undergraduate mathematics (Maclaren, in Faggiano & Montone, 2013 and Marshall, Buteau & Muller, in Faggiano & Montone, 2013),
- courses for prospective teachers of mathematics (Abu-Elwan, in Joubert, Clark-Wilson, & McCabe, 2011; Gurevich & Gorev, in Amado & Carreira, 2015);
- courses for practising teachers/lecturers of mathematics (Aldon et al., in Faggiano & Montone, 2013; Clark-Wilson, in Amado & Carreira, 2015 and Thurm, Klinger, & Barzel, in Amado & Carreira, 2015).

We predict that in the coming years, the ICTMT conference series will offer a fertile ground within which new acorns can germinate.

References

Adu-Gyamfi, K. (1993). *External multiple representations in mathematics teaching*. A thesis submitted to the Graduate Faculty of North Carolina State University in partial fulfillment of the requirements for the Degree of Master of Science.

Amado, N., & Carreira, S. (Eds.). (2015). *Proceedings of the 12th International Conference on Technology in Mathematics Teaching*. Faro, Portugal: University of Algarve.

Artigue, M. (2002). Learning mathematics in a CAS environment: The genesis of a reflection about instrumentation and the dialectics between technical and conceptual work. *International*

Journal of Computers for Mathematical Learning, 7(3), 245–274. doi:10.1023/A:1022103903080

Bardini, C. Fortin, P. Oldknow, A., & Vagost D. (Eds.). (2009). *Proceedings of the 9th International Conference on Technology in Mathematics Teaching*. Metz, France.

Bartolini Bussi, M. G., & Mariotti, M. A. (2008). Semiotic mediation in the mathematics classroom: Artifacts and signs after a Vygotskian perspective. In L. English, M. Bartolini Bussi, G. Jones, R. Lesh, & D. Tirosh (Eds.), *Handbook of International Research in Mathematics Education* (2nd ed., pp. 746–805). Mahwah, NJ: Lawrence Erlbaum.

Borba, M. C., & Confrey, J. (1996). A student's construction of transformation of functions in a multiple representational environment. *Educational Studies in Mathematics, 31*(3), 319–397.

Borovcnik, M, & Kautschlitsch, H (Eds.). (2002a). *5th International Conference on Technology in Mathematics Teaching: Plenary lectures and strands*. Klagenfurt, Austria.

Borovcnik, M., & Kautschlitsch, H. (Eds.). (2002b). *5th International Conference on Technology in Mathematics Teaching: Special groups and working groups*. Klagenfurt, Austria.

de Freitas, E., & Sinclair, N. (2013). New materialist ontologies in mathematics education: The body in/of mathematics. *Educational Studies in Mathematics, 83*(3), 453–470.

diSessa, A., & Cobb, P. (2004). Ontological innovation and the role of theory in design experiments. *Journal of the Learning Sciences, 13*(1), 77–103.

Dreyfus, T. (1991). On the status of visual reasoning in mathematics and mathematics education. In F. Furinghetti (Ed.), *Proceedings of the 15th Conference of the International Group for the Psychology of Mathematics Education* (Vol. 1, pp. 33–48). Genova, Italy: University of Genova.

Faggiano, E., & Montone, A. (Eds.). (2013). *Proceedings of the 11th International Conference on Technology in Mathematics Teaching*. Bari, Italy: University of Bari.

Fey, J. T. (1989). Technology and mathematics education: A survey of recent developments and important problems. *Educational Studies in Mathematics, 20*(3), 237–272.

Fischer, G. (2001). External and shareable artifacts as opportunities for social creativity in communities of interest. In J. S. Gero and M. L. Maher (Eds.), *Proceedings of the Fifth International Conference on Computational and Cognitive Models of Creative Design* (pp. 67–89). Sydney: University of Sydney.

Fraunholz, W. (Ed.). (1997). *Proceedings of the 3rd International Conference on Technology in Mathematics Teaching*. Koblenz, Germany: University of Koblenz.

Guin, D., & Trouche, L. (1999). The complex process of converting tools into mathematical instruments: The case of calculators. *International Journal of Computers for Mathematical Learning, 3*(3), 195–227.

Gutiérrez, A. (1996). Visualization in 3-dimensional geometry: In search of a framework. In L. Puig & A. Gutiérrez (Eds.), *Proceedings of the 20th conference of the international group for the psychology of mathematics education* (Vol. 1, pp. 3–19). Valencia: Universidad de Valencia.

Hoyles, C., & Lagrange, J.-B. (Eds.). (2009). *Mathematics education and technology—Rethinking the terrain: The 17th ICMI study*. Berlin: Springer.

Hoyles, C., & Noss, R. (2003). What can digital technologies take from and bring to research in mathematics education? In A. Bishop, M. Clements, C. Keitel, J. Kilpatrick, & F. Leung (Eds.), *Second international handbook of mathematics education*. Dordrecht: Kluwer Academic.

Jaworski, B. (Ed.). (1993). A bridge between teaching and learning. *Proceedings of the International Conference on Technology in Mathematics Teaching*. University of Birmingham, UK, LG Davis.

Joubert, M., Clark-Wilson, A., & McCabe, M. (Eds.). (2011). Enhancing mathematics education through technology. In *Proceedings of the 10th International Conference on Technology in Mathematics Teaching*. Portsmouth, UK: University of Portsmouth.

Kafai, Y. B., & Resnick, M. (1996). *Constructionism in practice: Designing, thinking, and learning in a digital world*. Mahwah, NJ: Lawrence Erlbaum Associates.

Kaput, J. (1999). Representations, inscriptions, descriptions and learning: A kaleidoscope of windows. *Journal of Mathematical Behavior (special issue), 17*(2), 265–281.

Laborde, C., & Sträßer, R. (2010). Place and use of new technology in the teaching of mathematics: ICMI activities in the past 25 years. *ZDM—The International Journal on Mathematics Education, 42*(1), 121–133.
Linn, M. C., & Peterson, A. C. (1985). Emergence and characterization of sex differences in spatial ability: A meta-analysis. *Child Development, 56,* 1479–1498.
Maull, W., & Sharp, J. (Eds.). (1999). *Proceedings of the 4th International Conference on Technology in Mathematics Teaching*. Plymouth, UK: University of Plymouth.
Milkova, E. (Ed.). (2007). *Proceedings of the 8th International Conference on Technology in Mathematics Teaching*. (CD-Rom) and online at http://fim.uhk.cz/ictmt8/seznam/?nazev=&autor=&typ=&v=1&Submit=Search (abstracts only).
Moreno-Armella, L., Hegedus, S., & Kaput, J. (2008). From static to dynamic mathematics: Historical and representational perspectives. *Educational Studies in Mathematics, 68,* 99–111.
National Council of Teachers of Mathematics. (1989). *Curriculum and evaluation standards for school mathematics*. Reston, VA: Author.
National Research Council. (1989). *Everybody counts*. Washington, DC: National Academy Press.
Noss, R., & Hoyles, C. (1996). *Windows on mathematical meanings: Learning cultures and computers*. Dordrecht: Kluwer Academic.
Olivero, F., & Sutherland, R. (Eds.). (2005). *Proceedings of the 7th International Conference on Technology in Mathematics Teaching* (Vols. 1 and 2). Bristol, UK: Bristol University.
Ozgun-Koca, S. A. (1998). *Students' use of representations in mathematics education.* In S. B. Berenson, et al. (Eds.), *Procedings of the Twentieth Annual Meeting of the North American Chapter of the International Group for the Psychology of Mathematics Education*, (Vol. 2, p. 812) Columbus, OH: ERIC Clearinghouse for Science, Mathematics, and Environmental Education.
Papert, S. (1980). *Mindstorms: Children, computers, and powerful ideas*. New York: Basic Books.
Park, D., Lee, J.-H., & Kim, S. (2011). Investigating the affective quality of interactivity by motion feedback in mobile touchscreen user interfaces. *International Journal of Human-Computer Studies, 69*(12), 839–853.
Ruthven, K., & Hennessy, S. (2002). A practitioner model for the use of computer-based tools and resources to support mathematics teaching and learning. *Educational Studies in Mathematics, 49,* 47–88.
Schwarz, B., Dreyfus, T., & Bruckheimer, M. (1990). A model of the function concept in a three-fold representation. *Computers & Education, 14*(3), 249–262.
Shaffer, D., & Kaput, J. (1999). Mathematics and virtual culture: An evolutionary perspective on technology and mathematics education. *Educational Studies in Mathematics, 37,* 97–119.
Skemp, R. (1978). Relational understanding and instrumental understanding. *Arithmetic Teacher, 26,* 9–15.
Tartre, L. A. (1990). Spatial orientation skill and mathematical problem solving. *Journal for Research in Mathematics Education, 21*(3), 216–229.
Triandafillidis, T., & Hatzikiriakou, K. (Eds.). (2003). *Proceedings of the 6th International Conference on Technology in Mathematics Teaching*. Volos, Greece: University of Thessally.
Verillon, P., & Rabardel, P. (1995). Cognition and artefacts: A contribution to the study of thought in relation to instrumented activity. *European Journal of Psychology of Education, 10*(1), 77–102.
Yeh, A., & Nason, R. (2004). Towards a semiotic framework for using technology in mathematics education: The case of learning 3D geometry. In *Proceedings of the 12th International Conference on Computers in Education*. Melbourne, Australia.
Yook, H. (2009). *A study on the types of interactive motions in Mobile touch interface (Doctoral dissertation)*. Korea, HK: Hongik University.

Appendix 1. The ICTMT Conference Series and Proceedings

ICTMT1—Birmingham, 1993. Jaworski, B. (Ed.). (1993). *A bridge between teaching and learning*. In *Proceedings of the International Conference on Technology in Mathematics Teaching*, University of Birmingham, UK, LG Davis.

ICTMT2—Edinburgh, 1995. Scott, T. (Ed.). (1995). *Informal Proceedings of the Second International Conference on Technology in Mathematics Teaching*. Edinburgh: Napier University.

ICTMT3—Koblenz, 1997. Fraunholz, W. (Ed.). (1997). *Proceedings of the 3rd International Conference on Technology in Mathematics Teaching* Koblenz, Germany: University of Koblenz

ICTMT4—Plymouth, 1999. Maull, W., & Sharp, J. (Eds.). (1999). *Proceedings of the 4th International Conference on Technology in Mathematics Teaching*. Plymouth, UK: University of Plymouth.

ICTMT5—Klagenfurt, 2001. Borovcnik, M, & Kautschlitsch, H (Eds.). (2002). *The 5th International Conference on Technology in Mathematics Teaching: Plenary lectures and strands*. Klagenfurt, Austria.

and Borovcnik, M, & Kautschlitsch, H (Eds.). (2002). *The 5th International Conference on Technology in Mathematics Teaching: Special groups and working groups*. Klagenfurt, Austria.

ICTMT6—Volos, 2003. Triandafillidis, T, & Hatzikiriakou, K. (Eds.). (2003). *Proceedings of the 6th International Conference on Technology in Mathematics Teaching*. Volos, Greece: University of Thessally.

ICTMT7—Bristol, 2005. Olivero, F., & Sutherland, R. (Eds.). (2005). *Proceedings of the 7th International Conference on Technology in Mathematics Teaching* (Vols. 1 and 2). Bristol, UK: Bristol University.

ICTMT8—Hradec Králové, 2007. Milkova, E. (Eds). (2007). *Proceedings of the 8th International Conference on Technology in Mathematics Teaching*. (CD-Rom) and online at http://fim.uhk.cz/ictmt8/seznam/?nazev=&autor=&typ=&v=1&Submit=Search (abstracts only).

ICTMT9—Metz, 2009. Bardini, C., Fortin, P., Oldknow, A., & Vagost D. (Eds.). (2009). *Proceedings of the 9th International Conference on Technology in Mathematics Teaching*. Metz, France.

ICTMT10—Portsmouth, 2011. Joubert, M., Clark-Wilson, A., & McCabe, M. (Eds.). (2011). *Enhancing mathematics education through technology. Proceedings of the 10th International Conference on Technology in Mathematics Teaching*. Portsmouth, UK: University of Portsmouth.

ICTMT11—Bari, 2013. Faggiano, E. & Montone, A. (Eds.). (2013). *Proceedings of the 11th International Conference on Technology in Mathematics Teaching*. Bari, Italy: University of Bari.

ICTMT12—Faro, 2015. Amado, N. & Carreira, S. (Eds.). (2015). *Proceedings of the 12th International Conference on Technology in Mathematics Teaching*. Faro, Portugal: University of Algarve.

Part II
New Spaces for Research

Returning to Ordinality in Early Number Sense: Neurological, Technological and Pedagogical Considerations

Nathalie Sinclair and Alf Coles

Abstract This chapter brings together recent research in neuroscience about the processing of number ability in the brain and new pedagogical approaches to the teaching and learning of number in order to highlight the significance roles of fingers and of ordinality in the development of early number sense. We use insights from these two domains to show how *TouchCounts*, a multitouch app designed for exploring counting and arithmetic, enables children to develop the symbol-symbol awareness that is characteristic of ordinality. We conclude by drawing out implications for further research making use of technology and neuroscience.

1 Introduction

In this chapter, we address two aspects of the use of technology within mathematics education. The first is what can be learnt from innovations in neuroscience in relation to studying the developing brain and the second is what can be gained from innovative touch screen learning technologies. The chapter, therefore, will address two issues:

- How does the use of brain mapping technology change our ways of thinking about and doing research?
- How can the use of technology support and foster innovative ways of learning?

In the first half, we draw on the work of Ian Lyons and others to trouble the typical developmental sequence posited by researchers of a movement from considering

N. Sinclair
Simon Fraser University, Burnaby, Canada
e-mail: nathsinc@sfu.ca

A. Coles (✉)
University of Bristol, Bristol, England
e-mail: Alf.Coles@bristol.ac.uk

E. Faggiano et al. (eds.), *Innovation and Technology Enhancing Mathematics Education*, Mathematics Education in the Digital Era 9,
DOI 10.1007/978-3-319-61488-5_3

actions on concrete objects, to the culmination in abstract mental structures. Innovations in brain technology mean we are now able to observe what takes place in the brain when faced with different tasks and placed in different contexts. A clear hypothesis emerging from this work is that what is significant in the learning of mathematics is not being able to link symbols to objects in a manner that is often considered accessible or natural, but being able to link symbols to other symbols. There is a danger that a focus on individual brains may lead to an impoverished and overly cognitive sense of what is important in learning. However, there are several neuroscientific studies that implicate the significant role of touch in the learning of early number and we consider two innovative research methodologies, inclusive materialism and enactivism, that take account of biology while not losing sight of the importance of the social and cultural in learning.

As an interlude, we then draw parallels between the suggestions coming from the neuroscience we review and the historical insights of one twentieth century educator (Caleb Gattegno). We suggest that Gattegno's innovative curriculum for early number can be seen as aimed at developing an awareness of symbols as relations and, by implication, developing an ordinal awareness of number.

In the second half of the chapter, drawing on our inclusive materialist/enactivist methodology, we analyse the use of the innovative *TouchCounts* iPad app, which supports children in the development of early number sense. Through a focus on sequences of interaction with the app, it appears that what students are becoming engaged and energised by is precisely the development of symbol-symbol awareness. We conclude by drawing out implications for further research making use of technology and neuroscience.

2 How Does Neuroscience Change Our Way of Thinking About and Doing Research?

For neuroscience researchers, the term *number sense* gets operationalized by looking for *neurological* changes that correlate to particular kinds of numerical tasks. More than a decade ago, Nieder and Miller (2003) believed they had identified the individual "number neuron" in rhesus monkeys. To do so, they presented the monkeys with pairs of slides containing dots on them that varied in size, shape or numerosity. By comparing the activation pattern in the monkey brains, they could identify the neurons that responded *only* to changes in numerosity. Even if there is validity to this suggestion, the situation seems to be much more complex for humans and identifying a number neuron is difficult because of the way that language, visual perspective and memory all play a role in number cognition. That said, neuroscience research has helped draw attention to the importance of underlying abilities such as finger gnosis and subitising, with subitising now being recommended practice in primary school classrooms (Clements, 1999). More recently, it has been shown that people who are successful at certain kinds of number tasks seem to draw on imagery associated with a number-line, which has helped shape some more visual approaches to early number (see Harvey, Klein, Petridou, & Dumoulin, 2013).

With these interesting insights coming out, it can sometimes be easy to forget the underlying assumptions that drive the associated neuroscience research. One such assumption is that cardinality should be first and foremost in the development of number sense (Butterworth, 1999). This assumption is evident both in the theoretical models that are proposed for how the brain processes number (see Bugden & Ansari, 2015), but also in the tasks that are used to study brain activity. Reigosa-Crespo and Castro (2015), for instance, focus on *magnitude processing* of numbers in symbolic and non-symbolic formats and describe "numerical magnitude" exclusively as cardinality: "To grasp the magnitude concept we need to learn the distinction between the transformations that do or do not modify the cardinality of a set (e.g., adding or removing objects in a set modifies the cardinality; spreading or grouping objects does not). We also need to compare between the numerosity of different sets (e.g., set A could be smaller, larger or equal to set B)" (p. 60). However, as Rips (2015) notes, there are many ways in which we use numbers in everyday life that do not depend on its cardinality, such as finding one's seat in the theatre and turning to a specific page in a book.

2.1 A New Space for Research on Early Number?

Recently, some researchers have challenged the dominant cardinal view of number cognition, and have proposed tasks that aim to engage ordinal thinking. Lyons and Beilock (2011) devised an ordinal task in which sequences of three numbers (or three sets of dots) were shown to participants, who then had to decide whether they were correctly ordered (either ascending or descending) or not. For example, the sequences [2, 3, 4] and [4, 3, 2] are correctly ordered but the sequence [2, 4, 3] is not. Speed and success on ordering tasks was strongly correlated to wider mathematical achievement. They argue that this experiment shows how a significant aspect of the meaning of a numeral, for students who are successful at mathematics, is relational and strongly tied to the unfolding of the *sequence* of numerals. There was a much smaller correlation with overall achievement, in the case of ordinal interpretations of dots. For this task, participants again have to judge whether three groups of dots are correctly ordered or not (e.g., from left to right, 2 dots, 4 dots, 5 dots are in order; 2 dots, 5 dots, 3 dots are not). A further distinction between tasks involving dots and numerals relates to the well established 'distance-effect' found in cardinal comparison of numbers, which is that the farther apart two numbers are, for (cardinal) comparison, the quicker it is found subjects are typically able to make judgements of which is bigger. Lyons and Beilock (2011) found the distance effect persisted with the order comparison of groups of dots but, importantly, when judging the order of numerals, the distance effect is reversed. In other words, when asked if three numerals are in order, the closer they are together, the quicker it is found that subjects can typically make the judgment or correct ordering or not. Lyons and Beilock (2011) use this reversal of the distance effect to suggest that the brain is doing something different when making ordinal comparisons of numerals, compared to both cardinal comparisons (of numerals or dots) and compared to ordinal comparisons of dots.

According to Lyons and Beilock then, the ordinal task on numerals calls upon the rote connections that we form in reciting the "number song". This hypothesis is interesting to consider in light of Seidenberg's (1962) theory of the ritual origins of counting, in which the recitation of the count list long precedes, historically speaking, the more cardinal counting of things (animals, people, money, etc.). Seidenberg argued that acts of ordinal counting are principally about calling forth the next or an(other), making the new or next appear, and not just about ordering that which is already visible. In research focused on developmental dyscalculia (DD), Rubinsten and Sury (2011) use the same task as Lyons and Beilock (both symbols and dots) with typically developing adults as well as adults with DD and found that both groups performed in similar ways on the symbolic tasks, but differed on the dot task.[1] They suggest that linguistic knowledge may facilitate ordinal number processing, which is consistent with Seidenberg's hypothesis and also revelatory of the difficulty of isolating any component of number sense, be it ordinal or cardinal.

Without adopting the reductive or innate assumptions underlying neuroscientic research, we find that the new emphasis on ordinality can be helpful to mathematics education research not only in highlighting the decisive role that task design plays in any definition (and assessment) of number sense, but also in drawing attention to aspects of number sense that have not been traditionally valued in the primary school curriculum. While still limited by the tools (fMRIs) used to identify brain response, which can only capture static images of brain activity, we see the new findings by Lyons and colleagues as drawing attention not only to the relational and symbolic aspects of number, but also to a powerful temporal sense of number. It is important also to note that fMRI images highlighting small regions of brain activity are only illustrating *differences* in brain activity (when provoked into doing a specific task compared to a control task). In other words, there are large regions of the brain also active, but what is flagged up are those regions that change activation patterns when presented with a number comparison task versus, say, a control reading task. It therefore does not seem to us entirely accurate to locate number processing in a particular brain region, or at least it must remain a possibility that there are vital elements of number processing taking place in distributed parts of the brain (which are also active when not doing number work).

[1]In the dot test, they varied the way that the dots were shown (such as changing the size of the dots) in order to see how perceptual cues modulated ordinal judgements. This is typical in the neurocognitive research literature on number sense.

2.2 A New Space for Research Methodologies?

The kind of neuroscience research that we have described above can be seen as challenging more sociocultural approaches to mathematics education research because it emphasises the biological component of mathematics cognition, sometimes even making claims about the innateness of certain aspects of number sense. Some of this research aligns well with embodied cognition theories, which stress the significant role that sensorimotor experiences play in mathematics understanding, though most mathematics educators do not address the neuronal level of embodiment. Sociocultural approaches, on the other hand, draw attention to the way in which the environment, language and politics are at play both in defining number sense, and in determining who is mathematically able.

The *inclusive materialism* developed by de Freitas and Sinclair (2014) provides a way of attending both to the sociocultural conditions of learning, while at the same time allowing a fundamental role for the body and the physical environment in mathematics teaching and learning. It does so by adopting a monist position that accords the same ontological status to mathematical concepts, human bodies, discourses and physical objects—this distinguishes it from theories such as instrumental genesis and semiotic mediation, which see tools and humans and concepts interacting with each other, but being ontologically distinct. A first consequence is that mathematical concepts are no longer seen as abstractions of sensorimotor experiences, but inevitably entangled with those experiences. In other words, a number is not a Platonic ideal, nor is it merely a sociocultural creation. Rather, it is an assemblage of counting fingers, things-to-be-counted, words to count with, and so on, that can be described as being in intra-action, which is Barad's (2007) term for describing the way in which concepts (such as number) are "specific material arrangements of experimental apparatuses" (p. 253). Barad's materialism is rooted in her interpretation of Bohr's work in quantum physics, where the concept in question might be light and the experimental apparatus might be the two-slit device that physicists use to study particular-wave duality. A concept such as light cannot be extracted from, or abstracted from the material assemblage in which it is measured; similarly, a mathematical concept such as number cannot be abstracted from the material assemblages in which it is encountered or used. In this perspective, the tool is not simply a way to mediate an already existing concept for the benefit of learning or understanding: the tool and the concept (and the learner) are in iterative entanglement. As Souriau (2015) would say, the tool and the concept and the learner do not exist in and of themselves, but in and of each other.

Bohr also helped problematize the taken-for-granted boundaries of the body when he wondered whether the body of a blind person holding a walking stick and navigating across a room stopped at the end of the person's hand, or at the end of the walking stick or at the edges of the walls. Similarly, does a learner's body extend to the screen on which she is manipulating shapes and symbols or contract to the Intraparietal Sulcus (IPS) [which is the brain region that lights up as people solve particular types of number-related tasks that require immediate cognitive judgements (Nieder & Dehaene, 2009)]? In the view of inclusive materialism, the idea of the human body as being well-defined by the contours of the skin melts away, since the body is both expanded to include physical tools and objects, as well as contracted, perhaps to a single neuron.

That a 'subject' is radically entwined with any 'object' that might be defined in contrast to it, and that any imagined demarcation between subject and object is malleable and porous, is an insight echoed in several other stances: a radical view of biology (Maturana & Varela, 1987); an approach to neuroscience (Varela, 1996); early thinking within cybernetics (Bateson, 1972); a strand of phenomenology (Merleau Ponty, 1962); and, within mathematics education, enactivism (see Reid & Mgombelo, 2015). Enactivism entails an essentially circular epistemology; what we take to be subject and object arise together and one cannot be taken as prior or more fundamental than the other. Perception does not represent a pre-given world to a subject, in fact it is misleading to even write of 'perception' as a noun; what we engage in, as humans, is "perceptually guided action" (Varela, 1999, p. 12), with touch providing a better exemplar than sight. It is clear that, in general, touch involves the activity of the 'subject' no less than that of the 'object'; to touch another human entails two decisions. The movement, choice and activity of the subject is perhaps harder to catch in the case of sight and hearing but, from an enactivist perspective, is no less present.

Common across both inclusive materialism and enactivism, therefore, is the view that an individual does not end at the boundary of the skin and that the social, cultural, political are enmeshed in the physicality of each of us and vice versa. If we are considering the learning of mathematics by children in school, we need to expand the typical view of the individual, to include relations that extend outside the body and, at the same time, inquire of larger systems, such as the social, cultural and political, how these make a difference to the relations and materiality that constitute each of us.

The novel theoretical approach described above requires new methodological considerations, since it would make little sense to individuate a priori the learner and the tool and the concept, and to see how the tool somehow causes the learner to "construct" an existing concept. In contrast to approaches that, for example, attempt to isolate and categories teacher knowledge (e.g., see, Chap. Studying the Practice of High School Mathematics Teachers in a Single Computer Setting) the research questions in inclusive materialism are more concerned with ontological issues, such as: what can the concept of number be within a particular assemblage (see Chap. Digital Mazes and Spatial Reasoning: Using Colour and Movement to Explore the 4th Dimension)? As with experimentation in quantum physics, we, as mathematics education researchers, might conceptualise a particular tool or apparatus as an experimental device that allows us to better understand the relations between matter and meaning that emerge in a particular classroom situation (de Freitas and Sinclair, 2016). Imagine, for example, as will be the case in this chapter, that the apparatus is an educational app. The apparatus would not simply be taken as a mediator of learning, or a tool that students use in order to learn a particular concept. Instead, the app is an apparatus that produces effects which help us see how meanings about the concept of number are entangled with the physical app and thus intrinsically indeterminate.

The next section is an interlude where we explore some historic pedagogical insights relevant to an ordinal approach to early number, following which we offer one example of what research within mathematics education might look like, from an inclusive materialist/enactivist perspective (illustrating our response to the question: how does the use of brain mapping technology change our ways of thinking about and doing research?). Our second aim is to show how neuroscientific results and pedagogical considerations pointing to the importance of ordinality in the early learning of number find resonance in the spontaneous activities of children working with a new technology (addressing the question: how can the use of technology support and foster innovative ways of learning?).

3 Returning to Ordinality: An Interlude of Pedagogical Considerations

We have found striking echoes of the neuroscientific suggestions, around developing symbol-symbol relations in the early learning of number, within the work of Caleb Gattegno, (1911–88). Gattegno worked across the world developing a mathematics curriculum based in the use of the Cuisenaire rods.[2]

Cuisenaire rods may appear to offer a strongly cardinal-based approach to number (and perhaps are often used by teachers in this way) with the white (1 cm^3) cube being associated with '1', the red cuboid with '2', the light green (length 3 cm) with '3' and so on up to '10'. However, as Coles (2014) has recently highlighted, this was far from Gattegno's vision. Rather than beginning with these quantitative relations, the first use of the rods, according to Gattegno (1957), was designed to foster more qualitative awareness ('greater than', 'less than') with letters for the colour names initially used to capture these relationships. Numerals are introduced to capture the special case where one kind of rod can fit a whole number of times into the length of another rod, see Fig. 1, where (if the longer rod is pink and the shorter ones are red) $p = 2r$.

In a context such as Fig. 1, numerals do not become associated with collections of objects (in a cardinal sense) but with a relation (in this case between lengths of rods). While '2' does not have a strong cardinal association when introduced as in Fig. 1, it does not have a strong ordinal sense either. Numerals are operators, or scale factors, and Gattegno's approach advocates the immediate introduction of rational numbers to indicate the inverse relation (Fig. 1 represents '½' as well as '2', $r = \frac{1}{2} p$). What quickly gets prioritised is the manner in which symbols link to

[2]Cuisenaire rods are cuboids with 1 cm^2 cross-sections and ranging in length from 1 to 10 cm. Each length is coloured uniquely (eg the cube with 1 cm lengths is white, the rod of length 2 cm is red).

Fig. 1 Introducing numbers with Cuisenaire rods

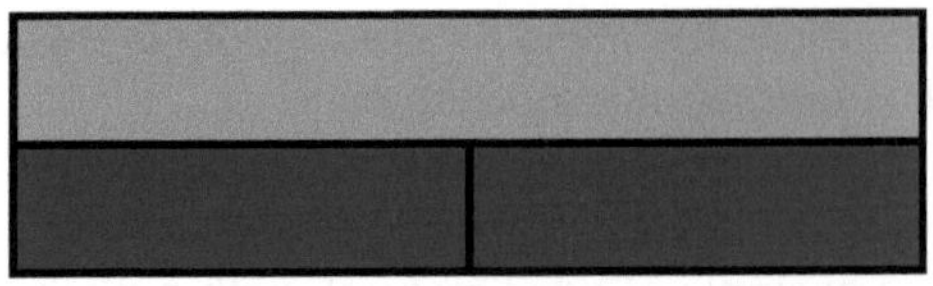

each other, for example the link between '2' and '½'. The materials are used to provide a context for symbol use but the number symbols do not represent the materials, they point to relations between the rods and attention can easily turn to relations between the symbols themselves (e.g., $t = 2p = 4r$, and $r = \frac{1}{4}t$). The link to ordinality, therefore, is that Gattegno's curriculum uses materials (the rods) as a mechanism to get children working on symbol-symbol relations, of which ordinality (the position of numbers in the number-count sequence) is one important aspect.

Gattegno uses a tens chart later in the curriculum (see Fig. 2). This chart can be used initially to work on number naming (the teacher might tap on '300', '50' and then '7', the children all chant back in unison 'three hundred and fifty seven'). The teacher might tap on a number and get children to chant back the number that is 1 greater, or 10 greater, or 100 greater, or 10 times bigger, or 10 times smaller, etc. What the chart makes available is the structure of our number system, not via thinking about the link between numerals and objects, but via the link between numerals and each other, how the names go together, and how you can get from one number to the next. In other words, the chart promotes ordinal awareness of number (see Coles, 2014 for further possibilities for working with children on the chart).

1	2	3	4	5	6	7	8	9
10	20	30	40	50	60	70	80	90
100	200	300	400	500	600	700	800	900
1,000	2,000	3,000	4,000	5,000	6,000	7,000	8,000	9,000
10,000	20,000	30,000	40,000	50,000	60,000	70,000	80,000	90,000
100,000	200,000	300,000	400,000	500,000	600,000	700,000	800,000	900,000

Fig. 2 Gattegno's tens chart

4 How Can the Use of Technology Support and Foster Innovative Ways of Learning?

In this section we draw on research, conducted within an inclusive materialist methodology, into the use of an innovative iPad app, *TouchCounts*. The data on children's use of this technology allows us to explore further the insights, discussed above, in relation to both early number and research methodology. This data has been drawn from a broader research project in which author Sinclair worked with children in two different daycare settings, one all-day daycare for three to five year old children and one after school day care for children five, six and seven years old. In both contexts, the research team interacted with groups of three to four children at a time, inviting them to engage in a variety of tasks that were designed both to take advantage of the functionalities of *TouchCounts* and to support their developing number sense. The app is 'radical' in the sense of Chap. Using Digital Environments to Address Students' Mathematical Learning Difficulties, meaning that the aim is to "propose fundamental mathematical content in innovative ways". There are no explicit 'tasks' or 'levels' built into the app and instead it offers a space for "being and becoming with others to make sense of the world … through mathematics" (Santi & Baccaglini-Frank, 2015, p. 122).

The data presented below was selected because it revealed different aspects of ordinality that we believe may be significant in children's learning. Our goal in selecting the examples was to exemplify new possibilities for working with ordinality. We analyse each example, in keeping with our overall research stance, through an ontological focus on what kind of number arises through the assemblage of students and apparatus.

TouchCounts was initially designed as a counting environment, to help children learn about one-to-one correspondence. Every time a finger touches the screen, a yellow disc appears, labelled with a numeral, and that numeral is spoken aloud. Each subsequent touch produces a yellow disc with the next biggest numeral on it. With the gravity mode turned on, taps that are made below the 'shelf' fall away, much in the same way that turning the page of a book makes that page number disappear. If one taps above the shelf, the yellow disc is 'caught' and remains on the shelf. It is thus possible to see just the yellow disk labelled '6' on the shelf if the previous five taps have been below the shelf. Indeed, putting just 6 on the shelf is a frequently-used task. Notice that it requires being aware of the fact that 5 comes before 6. Note that this task does not require any sense of cardinality. In both the temporal dimension, but also because of the lack of cardinal reference, this "Enumerating world" emphasises ordinality. With the use of the aural feedback as well as the numerals, there is also a strong emphasis on language and symbol, as per Lyons' recommendation.

The second world in *TouchCounts* has a more cardinal orientation. When three fingers touch the screen at once, a circle is created that contains the digit 3, as well as three smaller coloured circles. Once two or more such 'herds' are created, they can be added together through a pinching gesture. The resulting sum carries the trace of the colours found in the original herds. When the resulting herd is created, *TouchCounts* announces the sum out loud. There is a commonality with the e-pascaline software described in Chap. The Duo "Pascaline and e-Pascaline": An Example of Using Material and Digital Artefacts at Primary School in that particular gestures are associated with the processes of addition and subtraction. While the Operating World seems to focus activity around additive concepts, we will show in the examples below that significant aspects of ordinality are still at play. We begin, however, with two excerpts from the Enumerating World.

***Example 1**: Ordinal size* In a kindergarten classroom, the children are sitting on the carpet, with the overhead projector hooked up to *TouchCounts*. The teacher has asked the children to count by fives. They do this by tapping four times (simultaneously) below the shelf and then once above. This leaves the multiples of 5 on the shelf. The children take turns doing the 4 + 1 tapping. The following table gives a sense of the rhythm of the skip counting. Note that instead of hearing "five, ten, fifteen, …", the children hear "four, five, nine, ten, fourteen, fifteen, …".

Child's voice	Action	iPad voice
	Taps four fingers below the shelf	Four
Five	Taps one finger above the shelf	
		Five
	Taps four fingers below the shelf	Nine
Ten	Taps one finger above the shelf	
		Ten
	Taps four fingers below the shelf	Fourteen
Fifteen	Taps one finger above the shelf	
		Fifteen
	Taps four fingers below the shelf	Nineteen
Twenty	Taps one finger above the shelf	
		Twenty

The teacher had intended to only get up to about 25, but the children wanted to keep going. They eventually got to 100. At this point the shelf was full, so the teacher wiggled it, which made all the multiples of 5 fall down, off the screen. The children wanted to continue, and so the teacher let them do so, and 105 rested on the shelf, 110, 115, and so on. At 125, they began to predict what number would appear on the shelf—chanting it out, chorus style—and ended up going all the way to 200. At this point, the following interaction took place:

Cam: I thought that two hundred was right after one hundred, but it's not
Teacher: No, how far is it away from one hundred?
Cam: It's, it's, it's one more hundred away.

Analysis

Significant in this episode is the fact that the children were involved in a skip counting activity that had no explicit connection to a quantity of objects. The concept of number that emerges here is one of ritual acts, that is, a repeated gestural choreography that calls forth both names said aloud and symbols on the shelf (much like the ritual acts of counting that Seidenberg sees as the origins of counting). The concept of number is not a cardinal one. In addition, the highly patterned sequence of numbers on the shelf, 5, 10, 15, 20, 25, 30, etc., became predictable, enabling the children to chant out the multiples, ignoring each time the iPad voicing of the number immediately before the multiple of 5. The concept of number is temporal in that the sequence emerges over time, in time.

When 200 appears on the shelf, there had been no connection made between the number word and a quantity (of, say, two hundred objects). Indeed, Cam's realization about the relation between 200 and 100 is not a cardinal one. Instead, the concept of number that emerged in this assemblage was one in which getting to 100 involves the same time and sequence as getting from 100 to 200. In this sense, the relation seems to be deeply temporal, assembled as it is with the time it takes to create all the numbers up to 100 and then 200. The relation is also entangled with *TouchCounts*' pronouncements ("one hundred", "one hundred and forty-nine", "two hundred"), some of which these children may never have heard before, and which they could not have read from the symbolic forms (100, 149, 200). In other words, the meanings of the numbers were established by the way they were made (the gesture of four below, one above) and the time it took to get to them.

***Example 2**: Making 10* Three children are working together with Sean, a member of the research team. In prior visits to the daycare, the research team had noticed a tendency for the children to use just one finger when working in the Enumerating world, and wanted to encourage them to use multiple fingers at a time. The children had all succeeded in putting just 10 on the shelf. Now, Sean asked them if they can do it in a different way.

Immediately, Whyles touches

with 5 fingers (*TouchCounts* says "five"),
then 4 fingers (*TouchCounts* says "nine")

then 1 finger (*TouchCounts* says "ten") on the shelf.

The only thing on the screen is a yellow circle labelled 10 sitting on the shelf. The screen is reset and Benford goes next; he touches

with 5 fingers (*TouchCounts* says "five"),
then 1 finger (*TouchCounts* says "six"),
then 2 fingers (*TouchCounts* says "eight"),
then 1 finger (*TouchCounts* says "nine")

then puts 1 finger above the shelf (*TouchCounts* says "ten").

The only thing on the screen is a yellow circle labelled 10 sitting on the shelf. Auden went next. He touches

with 1 index finger (*TouchCounts* says "one")
then 1 pinkie finger (*TouchCounts* says "two")

then 1 index finger (*TouchCounts* says “three”)
then 1 middle finger (*TouchCounts* says “four”)
then 1 index finger (*TouchCounts* says “five”)
then 1 middle finger (*TouchCounts* says “six”)
then 1 index finger (*TouchCounts* says “seven”)
then 1 middle finger (*TouchCounts* says “eight”)
then 1 index finger (*TouchCounts* says “nine”)
then 1 middle finger (*TouchCounts* says “ten”)

There is nothing on the screen except the shelf. He presses reset and then touches

with 1 index finger (*TouchCounts* says “one”)
then 1 middle finger (*TouchCounts* says “two”)
then 1 index finger (*TouchCounts* says “three”)
then 1 pinkie finger (*TouchCounts* says “four”)
then 1 index finger (*TouchCounts* says “five”)
then 1 middle finger (*TouchCounts* says “six”)
then 1 index finger (*TouchCounts* says “seven”)
then 1 middle finger (*TouchCounts* says “eight”)
then 1 index finger (*TouchCounts* says “nine”)
then puts 1 index finger above the shelf (*TouchCounts* says “ten”)

Analysis

An equivalent activity of ‘making 10 in different ways’ could be done, for example, with a collection of objects, in which the children would be decomposing or ‘partitioning’ 10 into smaller quantities. In order to succeed at making different partitions of 10, children would need to pay attention to the sizes (the cardinality) of the separate groups they created and the overall numerosity. In transcribing what all three children did with *TouchCounts*, it can seem as though they ‘partitioned’ the number ten in a manner similar to that just described, the partitioning of 5, 4 and 1 for Wesley, of 5, 1, 2, 1, 1 for Benford and the full count (for Auden). But this is not the concept of number that emerges from this episode. Instead, 10 emerged as a number that you *get to*. You can get here quickly at first, and then slow down, like both Wesley and Benford, or you can get there steadily, one by one. In each case, getting to 10 involves first getting to 9, which suggests a privileged relation between 9 and 10, but the meanings of 9 and 10 are made through the shelf, which requires that 9 be placed below it before 10 can be placed above. And the relation becomes important only because 10 was set as the target. If Sean had asked for 11 on the shelf, a different privileged relation would have emerged: in this example though, 10 emerges in and of the shelf, the tapping, the task.

The two attempts by Auden suggest a different kind of intra-action in which the particular fingers—the index, the middle and the pinkie fingers—are entangled with the production of 10. In Auden’s first attempt to make ten, every other tap was done by his index finger and, except for one use of his pinkie, he alternated index-middle finger. This pattern of tapping took him ‘beyond’ the nine he needed to place below the shelf and he tapped below a tenth time with his middle finger. There is a sense here in which the production of numbers is essentially rhythmic, not just in the tap/iPad voice/yellow disc cycle, but in the middle finger/index finger cycle. But when "ten" is said aloud, and the disc falls of the bottom of the screen, Auden immediately tries again, seemingly aware of the fact that he had tapped too many times. On his second attempt, the alternating index finger tapping remained as a pattern up to nine and then he used the index finger again (the only time he tapped index-index in succession) to place his tenth tap above the shelf. In this way 10 emerged as an index finger number, the index finger being the one that begins and ends the sequence. We conjecture that making 9 or 11 would have been easier, since Auden's making of 10 involved a pause or disruption of a physical and temporal patterning.

Example 3*: Blurring the units* Transitioning out of an activity in which the children had been asked to put 10 on the shelf, Nathalie asked whether the children had played the "game" that involves tapping "one below, one above, one below, one above" the shelf in a sing-song voice. She touched five times towards the left of the screen and then Chleorah touched once below, once above towards the far right of the screen, and kept doing that while Olette said that it was her first time playing with *TouchCounts*. After resetting, Nathalie asked the children to watch her.

Screen action	iPad says	Others say
N touches below	one	
N touches above	two	
N touches below	three	
N touches above	four	
N touches below	five	
N touches above	six	A laughs
N touches below	seven	
N touches above	eight	
N touches below	nine	
N touches above	ten	N: You see that? Okay, can you read
		O: Two, four, six, eight, ten
		N: Two, four, six, eight ten. So what do you think is going to be the next number that shows up?
		O/C: Twelve
		N: How do you know that?

Chleorah explained that she knew how to count by twos. Olette then said very quickly “two, four, six, eight, ten, twelve, thirteen” and sat back and laughed. Nathalie said “ten, twelve, *fourteen*” with an emphasis on the last number and Auden said “fifteen”. Chleorah said “no, sixteen”. Nathalie asked the children to continue.

Screen action	iPad says	Others say
O touches below	eleven	
O touches above	twelve	
O touches below	thirteen	A: thirteen
O touches above	fourteen	A: fifteen
O touches below	fifteen	
O touches above	sixteen	(The shelf is full; O pulls her hand away and C takes over)
C touches below	seventeen	
C touches above	eighteen	A: eighteen (laughs)
C touches below	nineteen	
C touches above	twenty	
C touches below	twenty	
C touches above	twenty	
C touches below	twenty	
C touches above	twenty-four	
C touches below		A: twenty-one, twenty-two, twenty-fi, twenty-five
O/C touch above and below so quickly that the numbers are not said aloud		
O/C touch below and above a little slower, so that the iPad can be heard saying “thir” repeatedly		
C touches above	thirty-nine	
C touches below	forty	

Analysis

At the start of this sequence, Nathalie was focussed on a pattern of counting in twos. By tapping above and below the shelf it is possible to make only the evens appear above the shelf; the odd numbers literally fall out of view even though they have been said aloud by *TouchCounts*. Even and odd numbers are thus configured as ‘up’ and ‘down’ numbers, with the shelf acting as a kind of numberline on which consecutive evens can be placed. Odette’s pausing at 16, when the shelf is full, shows how the material arrangement of the discs and shelf briefly circumscribed the production of evens. But Chleorah picks up the up-down rhythm of tapping, filling up the shelf with overlapping discs and then speeds it up when she gets into the twenties so that only the tens digit can be heard. The up-down rhythm of the finger tapping configures the even numbers as limitless: if the children had counted by twos before, they had probably stopped at 20—given their age—but now that they do not have to produce the number names and their fingers can continue tapping, the even numbers extend beyond the familiar.

As the up-down tapping continued, the even numbers clutter up the shelf and the speed of tapping increasing, the soundscape changes from slow rhythmic counting to more repetitive "twenty" and then "thir" before alighting on "thirty-nine" and "forty". The echoing of Auden before (with "thirteen," "fourteen," "fifteen" as well as several numbers in the twenties) fades away. The four "twenty" in a row, being repetitive, may have provoked the speeding up of the tapping, with the sound or sight of thirty slowing it down just enough to be heard. Now, instead of the even and odd back and forth, the numbers bunch into batches that have a common initial sound—that is, they bundle into twenties and thirties, to the value of the tens digit. But the bundling is not only about the symbol in the tens place or the sound of the word, but also about time, that is, about the length of time it takes to go through the twenties and the length of time it takes to go through the thirties. A new pattern thus emerges in which the up-down binary of the tapping is joined with the bundling of the tens.

***Example 4**: Focussing on the digits* This episode, which lasts about thirty minutes, begins with one boy sitting in front of the iPad, creating and merging herds on the screen. He has made 155. Another boy, Henri arrives, and begins making herd as well, but not as expertly. The first boy Ned, goes away, and Henri continues working on merging herds. By the time Jordan and Dipak arrive, Henri has made 39. George says "thirty-nine Henri". He continues making his herd bigger and Dipak says, "Make one hundred. Henri, make one hundred." Ned returns and, having heard the comment, asserts, "I can do a hundred and fifty one." He goes away while Jordan and Dipak watch Henri work. Dipak says, "If you could make one hundred that would be awesome." Dipak laughs at a certain point when Henri has made 68 and says "six eight?" and Jordan says "sixty-eight". Henri makes 76 and he and George both echo *TouchCounts*, saying, "seventy-six". Dipak says, "Make a trillion, Jordan". Five minutes after they have begun, when Henri has made 80 (and there are other small herds on the screen) he says "okay, who wants to go next?" Jordan tries to put herds of 80 and 2 together but ends up making some new herds. He finally gets 82, and Dipak echoes *TouchCounts* and says "eighty-two". Henri says, "Let's make a trillion."

Jordan: Look how big this is (*see* Fig. 3a)
Dipak: Wooooow
Henri: What the heck? You need to use two fingers, not just one finger (*Jordan is trying to merge 88 and 1 but ends up creating new herds*)
Nathalie: What do you have there already?
Dipak: Two eights
Nathalie: Two eights?
Henri: Eighty-eight
Nathalie: Eighty-eight?

Fig. 3 Children making 100 on *TouchCounts*

Henri: (*looks up to the left*) And when there's two of the same number that's eighty-eight, forty-four, five, what is the number with two fives? (*Henri holds two fingers up, see* Fig. 3b)
Jordan: (*Jordan merges herds together to obtain 96*) Now look, sixty-f, sixty-six!
Nathalie: (*Responding to Henri*) Fifty-five
Dipak: Sixty-six?
Jordan: (*Makes a herd of 2 and merges it with the 96*)
iPad: Ninety-eight
Jordan: (*Makes another herd of 2 and merges it with 98)*
Henri: One hundred! (*clapping his hands while Nathalie takes Jordan's hands off the screen: see* Fig. 3c)
Ned: (*Coming back to the group*) I thought you were supposed to make a trillion
Henri: A trillion is after one hundred.

The boys continue making the herd bigger, now repeating the numbers after the iPad. They eventually make 204 and Dipal saying, "we got to 100 and then we made two hundred and four"). They make 208 and decide to show the iPad to others in the room. When they return to the table, and since *TouchCounts* has been reset, they start over again. They eventually, after about two minutes, manage to make 100 again, with Henri clapping his hands in excitement.

Analysis

The excerpt is replete with number-naming. *TouchCounts* names big numbers, the children repeat big numbers and also name numbers that they would like to make. Numbers are not quantities you count or operate on; they are discs/numerals/names that you can get to with sufficient creating and pinching of herds. We are struck by Henri's interest in the numeral 88 because we see in it the emergence of number as numeral not just to record or count or operation, but to consider as an object of interest in and of itself—in this case, an object that is special perhaps because of its twin-like character. Indeed, when 88 appears on the iPad, Henri muses aloud about the general situation of "when there's two numbers the same". He says 44 but gets stuck on 55 (which interestingly is 'irregular' in the sense that it could, and perhaps should, be named 'five-ty-five'). This focus on number-naming is ordinal in nature, in terms of its attention to symbols, but it also gives rise to yet another pattern (44, 55, 66, etc.) that almost completely ignores cardinality.

There is other evidence that the concept of number that emerges in this excerpt is concerned with the ordering of symbols associated with the herds the children have created. The children not only repeat the named numbers that *TouchCounts* speaks aloud but make statements about the order of the herds, as in when Dipak asserts that they made 100 and then 204 and when Henri asserts that a trillion comes *after* one hundred. We see it as significant that he uses the phrase "comes after" (an ordinal awareness) rather than, e.g., "is bigger than", in that number is once again about order rather than size.

Through the pinching of herds, number also emerges as something that you combine with other numbers, so that 6, for example, is not just the amount of objects on the screen, or the number after 5, but a device that can be used to make bigger numbers. It becomes a tool to make new numbers, including numbers as arbitrary as 204 and as magical as a trillion. Perhaps because they are working with large numbers, it is almost impossible for the children to actually attend to the quantity of coloured discs in the herd, which means that they are not attending to the relationship between the number symbol and the collection of disks. They are certainly attending to the growing size of the disc (at one point, George says that they should make the herd be bigger than the screen; and, at several other points, the boys talk about the size of the herd and their desire to make it bigger), which may support a more qualitative comparison between numbers—as in the comparison of 100 and 204 then a trillion and 100. The fact they are so motivated to get to 100 seems important in orienting attention towards number symbols, since these alone (or the *TouchCounts* voice) let them know if they reach their goal. Once again, number emerges from the (self-imposed) task, the possibility of endless creation and pinching of herds, echoing of large number-names and the presence of numerals.

5 Discussion

We began this chapter asking two questions: how does the use of brain mapping technology change our ways of thinking about and doing research? and, how can the use of technology support and foster innovative ways of learning? In this final section we will review what we have offered against these two areas of focus. Neuroscientific studies connecting brain mapping to items of behaviour seemingly draw attention back to an individual, constructing knowledge in isolation from others. Brain mapping technology points to the embodiment of mathematics, particularly in attempts to isolate particular brain regions with particular forms of mathematical activity. We see these technological advances as provoking a need for doing research in mathematics education that is thoroughly grounded in the material and that allows us to connect the materiality of the brain with the materiality of the world. Our inclusive materialist/enactivist stance allows just such an approach through a focus on the ontology of what arises as humans engage with apparatus and each other. Given our interest in the learning of number, one aim of the chapter has been to demonstrate how an analysis of interaction is possible that draws attention to the kind of number concept that arises.

Our review of brain mapping technology, as well as influencing the way we go about doing research, has also influenced our interest in ordinality in the early learning of number and, in particular, the way in which an innovative technology – the iPad app *TouchCounts*, fosters new ways of learning. A common approach to working on ordinality in schools involves practicing the number song; children are invited to count up to 5, then 10, then 20 and then 100. While we see much value in this practice, as a first way of introducing children to the language and sounds of number, much as in the ritual calling on to stage of deities thousands of years ago, we do not think it exhausts the potential of ordinal awareness. This has already been made evident in the work of Gattegno (1974), whose curriculum for early number was based on developing awareness of relations between lengths, where what are symbolised are relations between objects (greater than, less than, double, half), rather than, say, using numerals to label 'how many' objects are in a collection. Gattegno introduced work on place value, as a linguistic 'know-how' and not something that required 'understanding'. He made extensive use of fingers (both the teacher's and the children's) as haptic symbolic devices for working on number relations, with a focus on correspondence and complementarity. We see awareness of number, in this curriculum, arising out of linguistic skill and awareness of relations in a manner that does not emphasise a cardinal focus on counting collections.

We have argued that ordinality has a fundamental role in the successful early learning of number, firstly drawing on recent findings of neuroscience, which seem to support the important role accorded to language, symbol and relational understandings of number, and secondly identifying the ways in which number emerges from the particular intra-actions of the children's fingers, the spoken number names, the numerals, the shelf, the herds, the tasks, etc. In the examples we offered, we saw several instances of number relating, none of which seemed to be based on cardinal judgements, nor on explicit understanding of place value. We also saw evidence of the way the children attended both to the named words and to the symbols, even when the associated numbers are not part of their usual age-specific repertoire. We highlighted tasks that might usually be thought of as cardinal ones being offered in ways that invited ordinal thinking. These provide illustrations of the entanglement of concepts such as number with the experimental apparatus being used, the fingers being moved and the tasks being offered.

The concept of ordinality is often associated with Dedekind (1872/1963) who made use of the ordering of the continuum in order to define a real number in a rigorous manner. What we observe emerging in the examples above is a kind of ordering that is tied up with the touching, talking tool the children are using. The ordinality we observe from the assemblages of children and *TouchCounts* seems to take on a strong temporal aspect that may often be overlooked in writing-based contexts in which ordinals become static.

There are themes across the data on children's use of *TouchCounts*, in particular their apparent engagement and enjoyment and the creativity of their responses. In Example 1, the children want to take the count beyond 200 (compared to the teacher's plan of going up to 25). In the third example, there are several moments of laughter seemingly about the mathematics and perhaps the surprise of what *TouchCounts* says, and in Example 4 some of the excitement of the children can perhaps be read in the way they ask their own questions and pursue their own lines of thinking. It is worth considering why this might be happening—what it is about *TouchCounts*, the tasks offered to the children and what the adult does, that occasions such a response (which is typical in our experience)?

One answer to the questions above, that might be suggested by the neuroscience studies we reviewed, is that the context of the children's work in the four Examples allows a focus on number as a relation (see Coles, submitted). Whether the children are counting in 5 s, making 10, creating even numbers above the shelf, or making big numbers there is a consistent feature of their attention seemingly being drawn by pattern—patterns in time, in touch, in sound and patterns within the number name sequence. *TouchCounts* seems to free children from having to focus on cardinality (while not cutting off that important possibility) in order to explore the number system. Without the constraint of having to only work with numbers that can be grasped and handled (as would be implied by an approach to early number that emphasised the significance and difficulty of the move from concrete to abstract), the children become excited by exploring numbers and relationships far beyond what would be commonly expected at that age.

We conjecture that students of this age working with bigger than expected numbers and gaining awareness of number relations through playing with ordinality, will be developing quite different neurological structures to children offered a more typical cardinal-focused and concrete image of number. We see an important avenue of neuroscientific research (at the point where good enough brain imagining technology becomes wear-able and safe, perhaps) in exploring the brain effects of different pedagogical models of developing early number sense. At present it has been exclusively the case that neuroscientific studies related to education have led to implications for the classroom. In the spirit of the relational methodologies we use, we feel it is also time that principled classroom interventions lead to hypotheses to be tested neuroscientifically.

References

Barad, K. (2007). *Meeting the universe halfway: Quantum physics and the entanglement of matter and meaning*. Duke University Press.

Bateson, G. (1972). *Steps to an ecology of mind*. Chicago: University of Chicago Press, 2000.

Bugden, S., & Ansari, D. (2015). How can cognitive developmental neuroscience constrain our understanding of developmental dyscalculia? In S. Chinn (Ed.), *International handbook of dyscalculia and mathematical learning difficulties* (pp. 18–43). London: Routledge.

Butterworth, B. (1999). *The mathematical brain*. London: Macmillan.

Clements, D. (1999). Subitising: What is it? Why teach it? *Teaching Children Mathematics, 5*(7), 400–405.

Coles, A. (2014). Transitional devices. *For the Learning of Mathematics, 34*(2), 24–30.

Coles, A. (submitted). A relational view of mathematical concepts. In E. de Freitas, N. Sinclair, & A. Coles (Eds.), *What is a mathematical concept.* Cambridge: Cambridge University Press.

Dedekind, R. (1872/1963). *Essays on the theory of numbers*. New York, NY: Dover.

de Freitas, E., & Sinclair, N. (2014). *Mathematics and the body: Material entanglements in the mathematics classroom*. New York: Cambridge University Press.

de Freitas, E. & Sinclair, N. (2016). The cognitive labour of mathematics dis/ability: Neurocognitive approaches to number sense. *International Journal of Educational Research, 79*, 222–230.

Gattegno, C. (1957). *Numbers in colour: A new method of teaching the process of arithmetic to all levels of the primary school*. London: Heinemann.

Gattegno, C. (1974). *The common sense of teaching mathematics*. New York, NY: Educational Solutions.

Harvey, B., Klein, B., Petridou, N., & Dumoulin, S. (2013). Topographic representation of numerosity in the human parietal lobe. *Science, 341*(6150), 1123–1126.

Lyons, I., & Beilock, S. (2011). Numerical ordering ability mediates the relation between number-sense and arithmetic competence. *Cognition, 121*(2), 256–261.

Maturana, H., & Varela, F. (1987). *The tree of knowledge: The biological roots of human understanding*. Boston: Shambala.

Merleau-Ponty, M. (1962). *The phenomenology of perception*. London: Routledge & Kegan Paul.

Nieder, A., & Dehaene, S. (2009). Representation of number in the brain. *Annual Review of Neuroscience, 32*, 185–208.

Nieder, A., & Miller, E. K. (2003). Coding of cognitive magnitude: Compressed scaling of numerical information in the primate prefrontal cortex. *Neuron, 37*, 149–157.

Reid, D., & Mgombelo, J. (2015). Soots and key concepts in enactivist theory and methodology. *ZDM, The International Journal on Mathematics Education, 47*, 171–183.

Reigosa-Crespo, V., & Castro, D. (2015). Dots and digits: how do children process the numerical magnitude? Evidence from brain and behaviour. In S. Chinn (Ed.), *International Handbook of Dyscalculia and Mathematical Learning Difficulties* (pp. 60–77). London: Routledge.

Rips, L. (2015). Beliefs about the nature of number. In E. Davis & P. Davis (Eds.), *Mathematics, Substance and Surmise* (pp. 321–345). New York: Springer.

Rubinstein, O., & Sury, D. (2011). Processing ordinality and quantity: The case of developmental dyscalculia. *PLoS One, 6*(9) Epub.

Santi, G., & Baccaglini-Frank, A. (2015). Possible forms of generalization we can expect from students experiencing mathematical learning difficulties. *PNA, Revista de Investigacion en Didactica de la Mathematica, 9*(3), 217–243.

Seidenberg, A. (1962). The ritual origin of counting. *Archive for History of Exact Sciences, 2*(1), 1–40.

Souriau, É. (2015). The different modes of existence. (E. Beranek & T. Howles, Trans.). Univocal Publishing.

Varela, F. (1996). Neurophenomenology: A methodological remedy for the hard problem. *Journal of Consciousness Studies, 3*(4), 330–349.

Varela, F. (1999). *Ethical know-how: Action, wisdom and cognition*. Stanford, California: Stanford University Press.

The Coordinated Movements of a Learning Assemblage: Secondary School Students Exploring Wii Graphing Technology

Elizabeth de Freitas, Francesca Ferrara and Giulia Ferrari

Abstract This chapter uses assemblage theory to investigate how students engage with graphing technology to explore mathematical relationships. We use the term 'learning assemblage' to describe provisional dynamic physical arrangements involving humans and other bodies moving together and learning together. Emphasis on dynamic coordinated movements allows us to study how mathematics learning occurs in complex interaction with technology. We tap into the rich concept of 'sympathy' to understand the way that students develop a feeling for these coordinated movements as they participate collaboratively in mathematical investigations. Through sympathetic movements, a learning assemblage sustains a kind of *affective agreement* amongst the various bodies that participate. We show how assemblage theory helps us rethink the role of affect in technology tool use. This chapter sheds light on innovative ways of theorizing the role of Wii graphing technology in mathematical practice.

1 Introduction

In this chapter we explore how a sense of coordinated movement is entailed when students use Wii graph technology to explore mathematical relationships. We use assemblage theory and its *emphasis on relations between movements* in order to understand how these students are doing mathematics. The concept of *assemblage* has been taken up and used extensively in various new materialisms and new

E. de Freitas
Education and Social Research Institute, Manchester Metropolitan University, Manchester, England
e-mail: L.de-freitas@mmu.ac.uk

F. Ferrara (✉) · G. Ferrari
Università di Torino, Torino, Italy
e-mail: francesca.ferrara@unito.it

G. Ferrari
e-mail: giulia.ferrari@unito.it

E. Faggiano et al. (eds.), *Innovation and Technology Enhancing Mathematics Education*, Mathematics Education in the Digital Era 9,
DOI 10.1007/978-3-319-61488-5_4

empiricisms in the social sciences (Bennett, 2010; de Freitas, 2012; Fox, 2011; Mazzei, 2013). Much of this work follows assemblage theory as articulated by Gilles Deleuze and Feliz Guattari. According to this approach, assemblages are the fundamental "real unit" of study. Deleuze and Parnet (2007), for instance, claim that "the minimum real unit is not the word, the idea, the concept or the signifier, but the *assemblage*" (p. 51). In the inclusive materialist perspective of de Freitas and Sinclair (2014), the notion of assemblage is offered to de-essentialise the body and rethink its contours in mathematical activity, so that the *potentiality* of the body is stressed. Our focus in this chapter is on how human bodies collaborate and assemble with technology when exploring mathematical ideas. In another contribution in the book, Sinclair and Coles (Chap. Returning to Ordinality in Early Number Sense: Neurological, Technological and Pedagogical Considerations) draw on inclusive materialism to speak similarly of the role of the material environment in mathematics teaching and learning, shifting attention to the assembling of the human body with the concept of number. This chapter attempts to address some of the concerns raised about the concept of assemblage, particularly the concern that it is used all too often to simply name a *set* of individuals in a relationship (Buchanan, 2015). Indeed, we believe that the power of assemblage theory lies in the way it emphasizes how bodies are provisional relationships between *moving* parts, and that this *coordinated* movement involves an *affective bond* between components. In other words, a body is assembled through the dynamic force of affect, and not simply through mechanistic coordination of a set of components.

We use the term *learning assemblage* to describe provisional dynamic physical arrangements involving humans and other bodies *moving together* and *learning together*. Drawing on assemblage theory, we argue that perception occurs across these provisional arrangements and not in one central processing location (like the brain). This allows us to better study the way that acts of perception involve collaborative movement and activity. To "perceive" is actually to *assemble with* a relational environment in such a way as to fold back into it.

Such an approach highlights the concept of *proprioception*, which was originally defined as "sense of locomotion" and has evolved into the idea of "muscle sense" and a sense of one's own body's configuration. Proprioception is, by definition, a relational property of any assemblage. For instance, proprioception explains how we can move rapidly and without reflection in order to grasp a falling cup from a table. As one moves, the "proprioceptive potentialities" (p. 38) of the body are continuously reconfigured, as are the relative locations of objects in the foreground and background. This insight resonates with phenomenological approaches to perception, whereby *corporeal space* is "lived spatiality, oriented to a situation wherein the lived/living/lively body embarks on an architectural dance that actively spatializes (and temporalizes) through its movements, activities, and gestures." (Coole, 2010, p. 102). Proprioception is part of a larger concept called *kinesthesia*, which refers to the ability of the human body to feel its own movement and states, and thereby contributes to the sense that 'oneself' is the source of such action (Streeck, 2013). Sheets-Johnstone (2012) argues "not only is our perception of the world everywhere and always animated, but our movement is everywhere and always kinesthetically informed" (p. 113). These two ideas—proprioception and kinesthesia—are pivotal to our understanding of learning assemblages.

We use the concept of 'sympathy' to study the affective nature of coordinated movement in a learning assemblage. Although sympathy has various common sense meanings, we draw from the work of Gilles Deleuze who reclaimed the concept and tapped into its pre-Kantian meaning. For Deleuze, sympathy is a matter of independently moving bodies *moving together*, and involves the power of a body to affect and be affected (Deleuze & Parnet, 2007, p. 53). Sympathy thus takes on a pivotal role in understanding how affect is entailed in any learning assemblage. Following Deleuze and Parnet (2007), this ancient notion of sympathy, a term that comes from ancient Greek (sumpátheia) combines the meaning of "come together" and "pathos" and helps us understand how different bodies feel each other's movements. The notion of sympathy came to be used in diverse ways, but here we are interested in how it refers to a kind of agreement between bodies whereby they are mutually *affected* by each other through a coordinated movement. In our case, we want to study the way that affect plays a part in coordinated movements of different students insofar as they participate in a kind of collaborative and compassionate movement. Thus the learning assemblage is achieved insofar as affect sustains an agreement amongst the various bodies that join in. It is crucial that the term "agreement" not be interpreted as a judgement of rightness, but is rather a way of describing how bodies move together: "There is no judgment in sympathy, but agreements of convenience between bodies of all kinds" (Deleuze & Parnet, 2007, p. 52).

We next discuss a teaching experiment using Wii technology. We first present the context of research and the technology entailed, then the teaching experiment and finally a discussion of the data in terms of assemblage theory, drawing on the work of Manuel DeLanda (2006, 2011) who has further developed the ideas of Deleuze and Guattari.

2 The Context and Wii Technology

The experiment we present here took place in a secondary school in Northern Italy, as part of a wider study carried out during regular mathematics lessons. The study involved a class of grade 9 students participating in activities aimed at introducing the concept of function through a graphical approach using digital technology. The class was heterogeneously composed of 30 students (20 males and 10 females) from Torino and surroundings. The study lasted for 4 months and consisted of 9 meetings of 2 h in the period December 2014–March 2015. Two researchers (the second and the third author) designed and orchestrated the activities, while the teacher collaborated as an active observer within the classroom. The instructional methodology that was adopted offered diverse perspectives on the students' experience: collective discussions, group work, and individual work, often by means of written worksheets. The meetings took place in a laboratory room, which is used in the school as a laboratory for mathematical practice.

The activities were conceived so that the researchers could focus on kinaesthetic and proprioceptive experiences with tools that mobilized mathematical concepts related to functions. The approach draws on mathematics education research literature, which highlights that kinaesthesia and proprioception are part of mathematical understanding (see Nemirovsky, 2003; de Freitas, 2012, 2014; Ferrara & Ferrari, 2015; Roth, 2015). The teaching experiment focused on the spatio-temporal relationships that allow students to capture and describe motion phenomena, so it is greatly relevant for the study of mathematical functions, in part due to the historical roots of this particular area of mathematics, based as it was on the study of movement (e.g. Edwards, 1979). This research makes use of technology that is related to the game console Nintendo Wii because of the potential that it offers in terms of playing games through proprioception and kinaesthesia. The devices under consideration are the remote controllers (also called Wii Remotes, or Wiimotes) and the Balance Board of the Wii. The remote controllers are devices with which users can control and play games where real movement simulations are produced. The Wii balance board is usually used for games that depend on balance and body perception in space. So, bodily activity is crucial during activities performed with the Wii: the movement of the controller in the hand, the board under the feet, susceptible to all the variations in the player's balance, the eyes gazing at the feedback on the screen. The bodily actions required are kinaesthetic activities that deeply involve the proprioceptive capacities of the person who is participating. In a similar manner, Baccaglini-Frank and Robotti's contribution (Chap. Using Digital Environments to Address Students' Mathematical Learning Difficulties) discusses proprioceptive and kinaesthetic interactions with specific software as ways of accessing mathematical thinking by learners with disabilities.

The very first challenging step in this research project was to understand *how* to use the Wii devices in suitable pedagogical ways. Growing attention to *gamification* and *serious games* paradigms for education was seen as a way to tap student willingness to engage with the technology, which led to considerations of how to use the Wii as a resource for mathematics thinking and learning within a game context. Indeed, for many students such technology is already associated with game experiences, where players use the Wii technology to move through and solve problems within a virtual environment. Moreover, research suggests that affect might play a huge role in these kinds of game experiences. We were able to bring this technology to bear on pedagogical concerns through the use of two software applications, *WiiGraph* and *DarwiinRemote*, respectively working with two Wiimotes and one Balance Board. The first application has been developed with didactic goals by a group of researchers in mathematics education at the Centre of Research in Mathematics and Science Education of San Diego State University: Ricardo Nemirovsky and his colleagues (Nemirovsky, Bryant, & Meloney, 2012). The second software is freely available online. WiiGraph opened a wide range of opportunities to work with Cartesian graphs generated using Wii remotes. As the player moves her remote, the graph is depicted in real time on a single plane and captures instant by instant the movement of the corresponding controller. The graph on the screen documents the distance of the remote from an origin point, given by a sensor bar, which is positioned in the interactive space. Two players can play at the same time, and two different graphs can be shown on the screen.

Using a different modality, WiiGraph also allows working with a single graph which *assembles* the movements of the two remotes. This entails processing and integrating the two different movements in terms of one movement—that being the one graph that is collaboratively produced. These kinds of graphs lend themselves to two-person collaborative tasks involving two spatial variables, and can include activities of creating a rectangle or circle or some other figure, where one player controls the x-coordinate and another controls the y-coordinate. In a similar way, when connected to a Balance Board with a person standing on it, DarwiinRemote furnishes a dotted line on the screen. This line documents instant by instant the position of the person's centre of gravity, depicting the dotted graph of its projection on the horizontal plane (the plane of the board). This graph captures the horizontal motion trajectory of the centre of gravity. Beyond the original aim, the combined use of both software tools in the classroom allowed for rich explorations of the relations between motion laws and corresponding planar motion trajectories in the context of modelling motion. Even though we do not expand discussion about them here, design principles also play a role in our research in terms of novelty as for the studies presented in this book especially by Kynigos (Chap. Innovations Through Institutionalized Infrastructures: The Case of Dimitris, His Students and Constructionist Mathematics), Maschietto and Soury-Lavergne (Chap. The Duo "Pascaline and e-Pascaline": An Example of Using Material and Digital Artefacts at Primary School), and Tabach and Slutsky (Chap. Studying the Practice of High School Mathematics Teachers in a Single Computer Setting).

This chapter centres on a specific teaching experiment concerning sinusoidal functions and their relationships with a circular trajectory. In the experiment, the students made use of the first software, WiiGraph, to work with two types of graphs: *Line* graphs and *Versus* graphs.

3 WiiGraph Software: *Line* and *Versus* Graphs

WiiGraph is an interactive software application that takes advantage of the Wiimotes' multiple features to detect and graphically display the location of two users as they move along life-size number lines. In our experiment, an interactive whiteboard was also present in the classroom for visual experiences with the graphs projected by the computer screen, as well as a wide interaction space in the middle of the room to enhance students' opportunities for embodied and kinaesthetic experiences with the controllers.

A graphing session with WiiGraph commences when each user holds the controller pointed toward the sensor bar, so that a diffuse circle, matching a specific colour for each Wiimote, appears in the graph area. The diffuse circle is an index of the fact that the bar is capturing the distance of the controller at that moment—the circle indicates that the sensing technology and the software are coordinated. Once

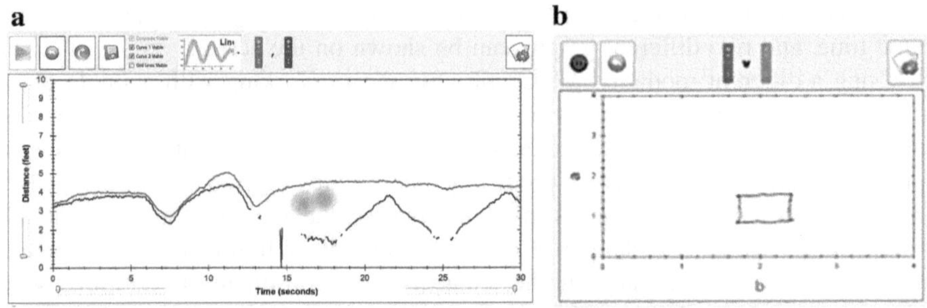

Fig. 1 **a** Graphs with *Line* modality. **b** Graphs with *Versus* modality

each circle is visible, WiiGraph can produce real time graphs with lines of the corresponding colour. Students can individually and collaboratively explore and work with several graph types, challenges, and composite operations, including shape tracing, maze traversal, and ratio resolution.

The graphs are configured in the graph area according to selected graph type, operations, and parameters like ranges, time periods and targets. Visibility controls can also be toggled during or after the session to selectively hide and show particular characteristics.

Among the various visual experiences that WiiGraph provides to the students, two are the most interesting graphical types for our study: *Line*, on the one hand, and *Versus*, on the other. We briefly discuss the two kinds of graphs to understand their functioning and the dynamic modes of interactions they offer to the students.

The *Line Graph* type, without target and operation, allows for depicting two distance-time lines of the kind $a(t)$ and $b(t)$, which correspond to the two Wiimotes' movement in front of the sensor bar, where a and b are the positions of the controllers and give their distances from the sensor. The thin coloured graphs are shown on the same Cartesian plane in the fixed time interval and they correspond to individual users (Fig. 1a). If an operation is selected, for example the sum $a + b$, this type adds to the previous lines a new distance-time line, which is the result of the operation at any given instant, in this case $(a + b)(t)$.

The other type, which is the very focus of our study, is the *Versus Graph* type. *Versus* plots an ordered pair of the distances of each user over time (the creation of the ordered pair is implicit). Briefly speaking, the graph that appears on the screen as result of the movement of the two Wiimotes is in this case the line $a(b)$, which is composed of the pairs of the kind $(b(t), a(t))$, for each t of the interval under consideration. The graph is thus always a spatial graph, where the variable of time disappears from the axes. In this perspective, one of the most significant challenges offered by *Versus* involves, as already said above, the creation of plane shapes, like rectangles, diamonds and circles (see e.g. Fig. 1b). Interestingly, this modality offers the students the opportunity of working together to collaborate and coordinate with each other for reaching a common goal.

4 The Teaching Experiment

In the teaching experiment discussed in this chapter, the students worked with WiiGraph in an activity focused on diagrams produced by the movement of two Wiimotes. At this point of the study, the students had already worked with WiiGraph on various aspects of functions through other activities using the *Line* modality: for example, they had explored plane transformations of graphs, such as translation and dilation; operations on functions, like the sum of functions; and relations amongst families of functions, like parallel straight lines, etc. They had also used the technology to face challenges that required matching suitable movements with given graphs (which offered room to reason on the role of the independent variable). Each of these activities required the students to use and compare with each other the two space-time graphs of $a(t)$ and $b(t)$. In this case, no constraint about coordinating the movements of the two Wiimotes in the interaction space was, implicitly or explicitly, given by the task. Rather, each learner could do a movement in a totally independent way with respect to the other learner. We discuss here a completely new activity, following these ones, in which pairs of students were asked to use the *Versus* modality of WiiGraph to generate together a single 'spatial graph' of the kind $a(b)$. These spatial graphs were the rectangle, the rhombus, and the circle (see again Fig. 1b). The novelty of the activity resides, at this point, in the fact that the students had to discuss ways of combining and coordinating the movements of the two controllers in order to produce one of these planar graphs. Concerning the rectangle and the rhombus, the students' first explored trials with the controllers. As their discussions evolved, they talked about how the changing positions of the two controllers were connected and assembled in the plane figure that they were seeing on the screen, referring to these positions as horizontal and vertical components of movement. Next, the task aimed to make this idea of horizontal and vertical components explicit for the whole class, asking two students, in front of their mates, to imagine being these orthogonal components on an imaginary vertical plane in space. The second author stood in front of them and gestured in space the plane figure (rectangle, rhombus), by moving her hand, while the two students had to move their right hands miming simultaneously the movements of the two components (see Fig. 2a for the case of the circle). This was purely a gestural affair, without the Wiimotes, in order to first explore the kinds of movements entailed in this task. The task involves the two learners in commencing their movements together with the researcher, synchronizing so that the learners' hands are always (1) at the same distance above the floor and (2) at the same width from the wall, as the researcher's hand. In so doing, the researcher's movement dictates the timing of the students' hand movements and the way in which they have to be assembled so that the original figure is the combined effect of such movements.

Then, the teaching experiment had the students return to the use of WiiGraph for obtaining a specific plane figure, taking advantage of the previous experience. Unlike the orthogonal gestures in the previous miming experience, the WiiGraph

a b c

Fig. 2 **a** Circle and components' movements in the air. **b**, **c** Lucrezia and Barbara coordinating the Wiimotes' movements

technology requires the two movements of the x and y coordinates (b and a, respectively) be performed *in parallel*, adding another complex dimension to the task. The WiiGraph software demands that the two controllers are moved along parallel lines in front of the sensor bar, although the sensory data is then processed as orthogonal. The movements, which before were orthogonally driven by the researcher's hand movement, and therefore more appropriately linked to habitual ways of characterizing these two components of planar movement, now need to be assembled in a different way. Indeed, the linear movements of the two Wiimotes have to occur along parallel lines, even though their combined effect will produce a similar two dimensional planar figure on the screen. In addition, the students need to agree with each other, driven by the software's feedback, so that the movements occur in suitable timing with each other and at suitable but different speeds for the combined movement to achieve the specific figure.

Thus the task entails tapping into time or duration in challenging ways, combining individual heterogeneous rhythms to achieve a third definitive rhythm. This third rhythm is then expressed as the target shape—be it the rectangle or rhombus. This task points to the fundamental role of time or duration in theories of embodied mathematics. It also shows how the learning assemblage implicates a confluence or commingling (like an orchestra or river) of diverse speeds and movements.

In this chapter, we focus on an excerpt from the video data collected during this teaching experiment when the students turned to the task of how to produce a circle using *Versus* and two Wiimotes. Two students, Lucrezia and Barbara, came in front of the class and mimed the gestures of the second author (Fig. 2a). They tried to explore the hand movements that might be needed to generate an imaginary circular trajectory in the air. Then, Lucrezia and Barbara started moving the Wiimotes with the aim of producing a circle as a third movement on the screen. Figs. 2b and c show the two girls while they are moving the controllers trying to be synchronized both in rhythm and speed. Fig. 3a shows the new circular movement that they are able to obtain on the screen (after some trials).

In order to better grasp the ways in which these coordinated movements of these two girls relate to the mathematics of the software and the mathematics of the figures they are making, we describe here the specific quantitative relationships that

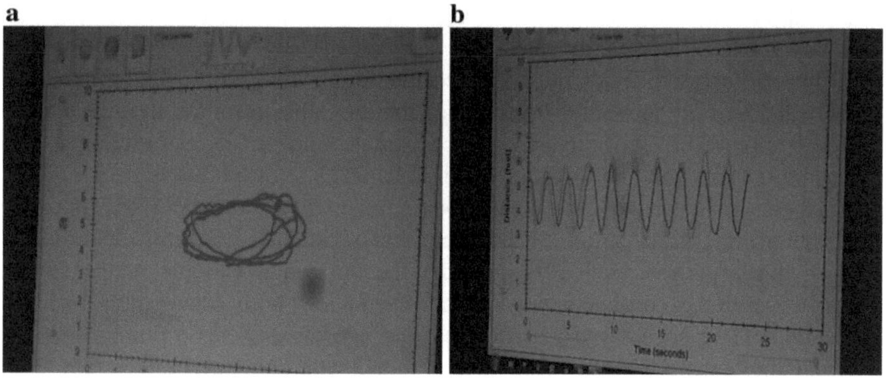

Fig. 3 **a** The circle with *Versus*. **b** The two periodic functions with *Line*

are at work in the *Versus* mode in which they are working. This is the mode where the time variable is implicit, and the two coordinated movements must fuse to make the figure, in this case either a rectangle, rhombus or circle. The diagram of these figures, belonging to the plane *ba*, is constituted of pairs (b, a), which capture specific positions in the plane and corresponding specific locations of the Wii remotes. Each position depends on the distances of the two controllers from the sensor during movement. If (b_1, a_1) and (b_2, a_2) are two distinct points of the diagram, they differ from each other in terms of movement, for the times at which the controllers reach the corresponding distances, say t_1 and t_2. So even if time seems to disappear in the *Versus* modality, it is obviously crucial for the creation of the diagrams, since these depend on movement. Indeed, what marks this modality as crucial for the purposes of this chapter is that the movements of the two users need to be coordinated to obtain the specific shape.

For example, in the case of a square or a rectangle (with sides parallel to the Cartesian axes), each side is created by one controller staying still and the other moving at a constant speed. In the case of the horizontal sides, the user at distance b (horizontal coordinate) is the only one moving, while a vertical side requires just the user at distance a to move. Thus, instant by instant, speed is not the same for the two movements: while the one user keeps null speed, the speed of the other is to be different from zero. If we take another quadrilateral (like a non-squared rhombus or a slanted rectangle), the two movements imply again two different instantaneous speeds, but there is also the fact that the ratio between speeds is always constant. Finally, when the expected diagram is a circle, the coordination between the controllers is much more difficult, and introduces challenging ideas about the relationship of movement to resultant diagram. In the case of the circle, the need to produce a curved line makes the task different from all the previous ones that the students have engaged in. With the circle, the speeds still need to be different from each other, but the users have to move in varying speeds in the same unit of time: while one user is at maximum speed, the other is at minimum speed while having to change direction, and when one is accelerating, the other needs to decelerate, and vice versa. Most importantly, the ratio between the speeds now is not constant, requiring the users to modulate their accelerations. The two movements must be coordinated and in "agreement" insofar as they together form the desired figure.

The combined effect of the students' coordinated movement, assembling the two linear movements (one the horizontal variable, and the other the vertical variable) produces a non-linear movement—that being the circular movement that appears in the resultant 'real-time' graph of the circle. Each student's hand movement has its own rhythmic pattern, each acts as either the horizontal variable or the vertical variable, and in this case the hands together form a new body or assemblage, a third movement. Their speeds must be different but coordinated. In so doing, their two different movements become conjoined, producing a combined effect. The combined effect is produced through the shared timing of their movements, creating a new movement that assembles the circular, creating the non-linear out of the combinations of linear movements. Each hand movement has its own rhythmic pattern, and each hand must move at a different speed, and indeed at related rates of changed speed, in order to achieve the effect. Thus the two bodies are moving together but apart, and the coupling of these movements forms a third movement that belongs to neither of the original bodies.

This task involves the combined movement of the two girls and the WiiGraph diagram. The two girls look at the screen where the graph appears but also down at their two hands as they move back and forth. We see here how time is the medium by which these two different movements commingle, and that a shared time must be adopted in order to create a new curve. This shared synchronized time introduces a new dimension to the combined movement, and yet each of the two human bodies has to follow different rhythms and speeds to achieve the effect. Thus the task points to the fundamental role of time or duration in theories of embodied mathematics, underscoring the way that bodies are assemblages of speeds and movements, while problematizing how we typically understand a body. We see how there is bodily agreement or coordination that characterizes the process of assembling, an aspect of group formation or body formation that is often overlooked. Agreement, as we use it, does not mean identification amongst parts, nor the creation of a unified homogeneous assemblage, but is used here to describe the coordination of heterogeneous movements – for instance a symbiotic relationship entails an agreement between two very different bodies that move together in a productive assemblage without erasing their distinctness. It is not that the two girls form bonds because they identify with each other, but because they are to become *coordinated* together. In that sense, there is a strong spatial element involved in the affective bonding that we see in forming this learning assemblage.

In the final part of the activity under discussion, after achieving the circle in the *Versus* modality (Fig. 3a), the researchers changed the software from *Versus* to *Line* modality. This modality calls time back as the independent variable in the space-time graphs, which capture over time the movement of the two Wiimotes as the changing distance from the sensor. The same two girls worked the controllers, and were asked to continue the movements that produced the circle in the other modality. The researcher repeated again and again "continue, continue" but said nothing else. The technology does not allow for having both space-time graphs and the circle (the spatial graph) present at the same time on the screen. So, the students have now to re-assemble in imagination the new coordinated movement of the circle graph, but now they produce the graphs of two periodic functions. The periodic functions that appear on the screen are the effect of using the *Line* modality (see Fig. 3b). The periodic functions (one for each girl) repeat horizontal or vertical coordinates according to the winding of the circular movement, graphed against the time variable. The class now sees the two girls continue to make the same hand movements, but now instead of a circle they generate two sinusoidal graphs. The teaching experiment helps the students grasp the many different ways in which related movements are at work in the apparently fixed and familiar figure of the circle, deepening their understanding of the geometric figure, and enhancing their embodied understanding of the mathematics involved. The *Line* modality shows the wave function for each of these movements, and shows where they intersect, directing attention to when the movements must be in some sense 'equal'.

In unexpected ways, this series of activities has "closed" the circle. From the initial experiences, which involved the researcher's hand as a catalyst of rhythm and speed, to the coordinated movements that produced the circle graph, learners pass to this last motion experience, which makes present the rhythm and speed of time in the production of the circle. In other words, moving back and forth between the *Line* and *Versus* modality allows the students to see how the same movements generate a circle and two periodic functions, thereby bringing to agreement these two ways of thinking about the movements inherent in making shapes. The periodic graphs demonstrate and indeed emphasize *the coordinated rhythm* of the students' movements.

5 Discussion: Assemblage Theory

In this section, we use *assemblage* theory to analyse the kind of mathematics learning that occurs as students assemble with technology in mathematics classrooms. The experiment discussed in the previous section sheds light on *innovative* technologies for teaching and learning mathematics. Assemblage theory, however, helps us analyse this data less in terms of tool use and more in terms of the affective force of the technology, insofar as it participates in the learning assemblage. The focus on learning assemblages is new *per sé* in the panorama of mathematics education literature on technology, which has tended to study technology in terms of 'tool use' and the affordances *for* the human. Here we look beyond the human at the entire assemblage that incorporates various kinds of non-human agencies. It is important to note, however, that an assemblage is *not* merely a set of bodies collected as one. Thus, it is not simply a matter of a student and a computer being seen as a cyborg, as for example recognized by Borba and Villareal (2005) when they speak of knowledge production by collectives of "humans-with-media". These collectives are "the basis for an epistemology that focuses attention on how people know things in different ways with the introduction of different technologies." (p. 27). Although our approach shares much with endeavour, there is an important difference regarding the episto-ontological claims. These authors state that the human should be considered as the epistemological subject, as the basic unit for thinking and of analysis in the production of knowledge. They focus on knowledge production and retain a dialectical relationship between the human and the technology: "We claim that a new technology of intelligence results in a new collective that produces new knowledge, which is qualitatively different from the knowledge produced by other collectives." (p. 24). We do not want to abandon this insight, but we want to draw attention to the more than human ontological relationships entailed in such a collective. Our chapter aims to study human-technology interaction without treating the human as a 'user' of a 'tool', because such an approach tends to over-emphasize human will and agency, as seen in the French perspective of instrumental genesis (e.g. Artigue, 2002; Baron, Guin, & Trouche, 2007). In other words, we aim to rethink the nature of distributed agency across an assemblage, and this entails rethinking the very nature of "use-value" since the idea harbours particular assumptions about agency (see for example the discussion around instrumental orchestration presented in Thomas et al.'s contribution in the book—Chap. Innovative Uses of Digital Technology in Undergraduate Mathematics). Moreover, we want to explore how the assemblage is assembled, rather than start with pre-given ideas about an individual who mediates outside sources of knowledge, like in the case of semiotic mediation or representational infrastructure theories (see Bartolini Bussi & Mariotti, 2008; Hegedus & Moreno-Armella, 2008). As Chorney (2014) points out, all too often "when the focus is on the student and the tool interacting, a dualist approach has been adopted." (p. 60). So, assemblage theory gives us a new perspective that helps us study the *more than human* process of "becoming-together", whereby the Wii, the students and the circle are entangled in the relational movement that characterises mathematical activity.

In particular, assemblage theory furnishes innovative ways of analysing the classroom episode, focusing on the affective and ethical nature of material entanglement. In the episode presented, the learning assemblage and its movement gradually emerge through various coordinated body movements, first between Barbara and Lucrezia, who must move so as to *agree* with the circular trajectory actualized by the researcher. The coordinated movement of the task re-assembles the students' hands into one movement forming a new body, that which actualizes the sinusoidal functions on the screen. The assembling process entails an ongoing agreement amongst various movements, which allows the collaborative activity to achieve something: that being a series of graphs that express mathematical relationships. We see in this coordinated agreement a way of addressing the ethical aspects of such activities, in that the students gradually and quietly become sympathetic with each other. This sympathy or agreement amongst various movements entails an ethical obligation to get the task done, or, following Barad (2010), shows us how "entanglements are relations of obligation—being bound to the other—enfolded traces of othering" (p. 265). The fact that bodies are related in terms of coordinated movements, and the fact that a graph is produced, relates directly to the force of affect that sustains the assemblage. Affect is the force or glue that sustains the sympathetic relations of agreement between Barbara and Lucrezia. Although it is difficult to track the evidence of this force, doing so helps us think about the ethical dimension of learning assemblages—there is an ethical obligation to the assemblage and its movement because affect glues the assemblage together (provisionally). There is a certain responsibility, which is part of the sympathetic manner in which the two girls are working together. If we think of this as a positive learning encounter, then this obligation to collaboratively engage with the task depends on the force of affect—engaging in the task depends on the fact that bodies have a capacity to be affected. Obligation is then a kind of coordinating with the other, a sympathetic agreeing with the other—not an identification, but a coordinated effort. The students affect each other. It is the power to be affected and to cause affects that produces the learning assemblage. For Deleuze, assembling *is* sympathy.

DeLanda (2006, 2011) will speak of assemblages as emergent entities within systems of matter, energy and information. The simplest assemblage, according to DeLanda, is formed when two molecular populations of air (or water) at different pressure or temperature are placed in contact. Because of the difference, a gradient is formed. This gradient is the simplest assemblage, having a tendency to dissipate but also a capacity to be exercised. Note that DeLanda (2011) defines assemblages in terms of a mathematical concept—the gradient—which is the derivative of a multi-variable function. A gradient is a vector whose components are the partial derivatives of an n-dimensional function. In this way, DeLanda operationalizes and makes more concrete the proposal that assemblages are relations of speed and movement. He is not simply suggesting that gradient is a good metaphor for how complex assemblages are formed. He is literally suggesting that assemblages are differentiation processes and relations of difference. As in our case study, the assembling of girls and technology is a gradient (a series of relations of speed and direction). Quite explicitly, the technology entails that Lucrezia and Barbara's instantaneous speeds are captured as the two derivatives $\mathrm{d}b/\mathrm{d}t$ and $\mathrm{d}a/\mathrm{d}t$ that constitute the speed of movement along the circle, as well as that their directions make the direction of this movement. Briefly speaking, WiiGraph assembles the derivatives in the slope of the line tangent to the circle point by point. This case study with the circle illustrates DeLanda's argument about how *assemblages are gradients*—quite literally, the study directs our attention to how the concept of circle is a series of speeds coordinated through the students' movements. Their timed accelerated movements *are* the gradients that are imperceptible in the graph. The assemblage of graph-concept-student is achieved through these gradients.

Our case study illustrates this point well because the graphs on the screen are produced only by a series of coordinated movements, and thus the achievement of a circle graph is in fact nothing more than a complex set of differentials (degrees of difference). What we take to be the unity of the achievement is actually a field of movement and differential relations of speeds and directions. All of these components partake of the mathematical work in the classroom; the mathematical work is determined by the relations between the components. For example, when the hand movements of two students are assembled with each other, an entirely new kind of movement emerges—a circular and also periodic movement emerges from the combining of two linear movements. Thus mathematical concepts of linearity, periodicity, etc. are at play in the emergence of this particular assemblage. The speeds of the movements are intrinsically relevant to the resultant properties of this assemblage—that being the circle graph and later the sinusoidal graphs. The circle emerges as a specific varying relationship of speed and direction, and the sinusoidal as a specific relation of varying slopes, both graphs emerging from the modulation of Lucrezia and Barbara's accelerations. The case study demonstrates how the radically new—in this case circular movement—can emerge from a set of components that are different in kind (linear). The linear can be combined to create the circular. We often take this for granted, but it is actually a philosophically significant action. The fact that the circular motion emerges in this way is a perfect example of how an emergent property can be distinctive and not possessed by the components of the assemblage.

It is important to understand that the identity of an assemblage is both embodied and *expressed* in its materiality. In other words, an assemblage is associated with a body (an individualized collective) *and* expressed information as "raw physical pattern." (DeLanda, 2011, p. 200). This dual emphasis on embodiment and expression is crucial for any theory of assemblage that aims to attend to learning. A learning assemblage must thrive through its gradient while also *express information*. A gradient in physics can be live or dissipated, and in either case it possesses the same energy. However, the live gradient expresses more information because it is ordered. A dissipated gradient lacks order. A high degree of information is associated with a highly structured assemblage. We can see in our case study how the assemblage of students and Wii technology is highly structured insofar as the students have been asked to do something and they are attempting to do it. Like any classroom, there is some authority that structures the activity. But the assemblage is also highly structured in a more systems-based way insofar as the various movements of components—the two different students, the software, the sensor—are carefully coordinated so that they can create the desired graph on the screen. For example, Lucrezia and Barbara cannot see any mark on the screen if they do not keep the controllers pointed at the sensor. Coordination between the students' movements and the sensor requires such a careful pointing during the motion experience. In a similar way, the girls need the software feedback to grasp the efficiency of their movements. They even do not know a priori that their movements need to respect the conditions discussed above. It is only when their coordinated movement realizes and sustains these conditions that they are able to assemble with the technology in a sympathetic way that they are then able to draw a circle-like diagram. It is precisely at this point that they experience the shared obligation to each other that Barad cites above, that sense of shared commitment and affective bond that achieves the coordinated rhythms and the circle graph.

The repeated attempts of the girls to achieve the graph shows how sympathies proliferate in everyday minute interactions, lived in and as affective bonds, and assemble into larger overt coordinated emotional responses between bodies. Minute sympathetic movements contribute to passionate attachments, so that the emotional investment in such shared activities becomes pronounced: "sympathy is bodies who love or hate each other, each time with populations in play, in these bodies or on these bodies." (Deleuze & Parnet, 2007, p. 52). Thus affect circulates across minute movements as the two girls coordinate their activity. We see the learning assemblage evolve through these relations, where sympathy becomes "something to be reckoned with, a bodily struggle". The girls do not identify with each other or 'put oneself in the other's shoes', but they assemble with each other and with the Wii, and thereby enter a process of *becoming other* that does not erase the other (Deleuze & Parnet, 2007, p. 53). An ethical relationship emerges through the sympathetic coordination of movement.

Thus there is an increasingly intense obligation amongst the many components of the assemblage as the teaching experiment unfolds. The assemblage is embodied in the relations between these participants, but is also expressed through *and in* the information in the graph. That information is precisely the mathematical relationships captured in the graph. There are many different ways in which the students might have made some marks on the screen using the WiiGraph technology, and all these different ways express different degrees of information. The desired outcome —the circle—is clearly considered as that which possesses more information by the adults present. Thus the sense of obligation that an entanglement entails, this sense that we are entangled together in the shared task, is both an affective sympathetic bond and an expression of information deemed to be information by the researchers who are present. DeLanda is careful to situate assemblages in historical and cultural context, and to recognize the contingency of how particular arrangements and movements are deemed to possess more information than others. Although we do not have the space to develop these ideas here, his approach is not a neutral or ahistorical theory of assemblages, but rather one that brings together insights from systems theory (in particular physical-chemical processes at work in systems) with affect theory and information theory, in such a way that the historical-political context is also integrated.

Summarizing, in this chapter we have used assemblage theory to offer new ways of examining the processes through which individual human bodies come together with technology in mathematical activity. Our innovative analysis highlights the *relational movement* that characterises the *entanglement* of the technology, the students and the mathematical concepts. The emphasis on relations between movements has allowed us to study how the entire learning assemblage is doing mathematics and thinking mathematically. With our episode we have proposed to look at the process of becoming-together as more than human, instead of focusing on the human as a user of tools. We have also seen how we have been able to frame the discussion of the classroom episode shedding light on the degree of obligation that is entailed in any assemblage, and how this helps us begin to think about the ethical dimension of learning, where questions of obligation and responsibility must be considered. This has brought us to draw attention to the role of affect in innovative technology and to think of it as the force that sustains a sympathetic human-technology assemblage. Affect is intended in Deleuzian terms, as the force that nourishes sympathetic relations of agreement between bodies whereby they are mutually affected by each other. Affect thus circulates across bodily relations of sympathy. Sympathy takes on a pivotal role in any learning assemblage. Accordingly, it is not a matter of identification but of coordinated effort of *agreeing* with the other—this agreeing allows for radically diverse forms of heterogeneous movement, and is not a matter of compliance or becoming the same. It is rather an attempt to think about how we form assemblages of radically heterogeneous movements in ways that are productive of learning and ethical relationships. In our case study, the bodily agreement or coordination produces rich mathematical thinking—an assembling of gradients and directions that speaks directly to the shape of the sinusoidal functions and their relationship with the figure of the circle. The learning assemblage that we have analysed here is a complex entanglement of affect and information, demonstrating how innovative technologies add to our understanding of fundamental aspects of mathematics learning.

References

Artigue, M. (2002). Learning mathematics in a CAS environment: The genesis of a reflection about instrumentation and the dialectics between technical and conceptual work. *International Journal of Computers for Mathematical Learning, 7,* 245–274.

Barad, K. (2010). Quantum entanglements and hauntological relations of inheritance: Dis/continuities, spaceTime enfoldings, and Justice-to-Come. *Derrida Today, 3*(2), 240–68.

Baron, M., Guin, D., & Trouche, L. (2007). *Environnements informatisés et ressources numériques pour l'apprentissage: conception et usages, regards croisés.* Paris, France: Hermès.

Bartolini Bussi, M.G., & Mariotti, M.A. (2008). Semiotic mediation in the mathematics classroom: Artifacts and signs after a Vygotskian perspective. In L. D. English, M. Bartolini Bussi, G. Jones, R. Lesh, & D. Tirosh (Eds.), *Handbook of International Research in Mathematics Education* (pp. 746–783). Mahwah, NJ: Lawrence Erlbaum.

Bennett, J. (2010). *Vibrant matter: A political ecology of things.* Durham, NC: Duke University Press.

Borba, M. C., & Villareal, M. E. (2005). *Humans-with-media and the reorganization of mathematical thinking: Information and communication technologies, modeling, visualization and experimentation.* New York, NY: Springer.

Buchanan, I. (2015). Assemblage theory and its discontents. *Deleuze Studies, 9*(3), 382–392.

Chorney, S. (2014). From agency to narrative: Tools in mathematical learning. *Ph.D. Dissertation.* Burnaby, BC: Simon Fraser University.

Coole, (2010). The inertia of matter and the generativity of flesh. In D. H. Coole & S. Frost (Eds.), *New materialisms: Ontology, Agency, and Politics* (pp. 92–115). Durham, NC: Duke University Press.

de Freitas, E. (2012). The diagram as story: Unfolding the event-structure of the mathematical diagram. *For the Learning of Mathematics, 32*(2), 27–33.

de Freitas, E. (2014). How theories of perception deploy the line: Reconfiguring students' bodies through topo-philosophy. *Educational Theory, 64*(3), 285–301.

de Freitas, E., & Sinclair, N. (2014). *Mathematics and the body: Material entanglements in the classroom.* New York, NY: Cambridge University Press.

DeLanda, M. (2006). *A new philosophy of society: Assemblage theory and social complexity.* London, UK & New York, NY: Continuum.

DeLanda, M. (2011). *Philosophy and simulation: The emergence of synthetic reason*. London, UK & New York, NY: Continuum.
Deleuze, G., & Parnet, C. (2007). *Dialogues II*. New York, NY: Columbia University Press.
Edwards, C. H. (1979). *The historical development of the calculus*. New York, NY: Springer.
Ferrara, F., & Ferrari, G. (2015). Becoming mathematical subjects by playing mathematical instruments: Gibbous lines with WiiGraph. In S. Carreira & N. Amado (Eds.), *Proceedings of the 12th International Conference on Technology in Mathematics Teaching* (pp. 378–386). Faro, Portugal: Faculty of Sciences and Technology of the University of Algarve.
Fox, N. J. (2011). Boundary objects, social meanings and the success of new technologies. *Sociology, 45*(1), 70–85.
Hegedus, S. J., & Moreno-Armella, L. (2008). Intersecting representation and communication infrastructures. *ZDM: The International Journal on Mathematics Education, 41*(4), 399–412.
Mazzei, L. (2013). Materialist mappings of knowing in being: researcher constituted in the production of knowledge. *Gender and Education, 25*(6), 776–785.
Nemirovsky, R. (2003). Three conjectures concerning the relationship between body activity and understanding mathematics. In N. A. Pateman, B. J. Dougherty, & J. T. Zilliox (Eds.), *Proceedings of the 27th conference of the international group for the psychology of mathematics education* (Vol. 1, pp. 103–135). Honolulu, HI: PME.
Nemirovsky, R., Bryant, C., & Meloney, M. (2012). *WiiGraph user guide. center for research in mathematics and science education*. San Diego, CA: San Diego State University.
Roth, W. M. (2015). Excess of graphical thinking: Movement, mathematics and flow. *For the Learning of Mathematics, 35*(1), 2–7.
Sheets-Johnstone, M. (2012). Movement and mirror neurons: A challenging and choice conversation. *Phenomenology and the Cognitive Sciences, 11*(3), 385–401.
Streeck, J. (2013). Interaction and the living body. *The Journal of Pragmatics, 46*, 69–90.

Using Digital Environments to Address Students' Mathematical Learning Difficulties

Elisabetta Robotti and Anna Baccaglini-Frank

Abstract The need to deal with different cognitive necessities of students in the mathematical classroom, and in particular of students who persistently fail in mathematics, frequently referred to as "having mathematical learning difficulties or disabilities" (MLD), has become an important topic of research in mathematics education and in cognitive psychology. Though frameworks for analyzing students' difficulties and/or for designing inclusive activities are still quite fragmentary, the literature rather consistently suggests that technology can support the learning of students with different learning characteristics. The focus of this chapter is on providing insight into this issue by proposing analyses of specific software with a double perspective. We will analyze design features of the selected software, based on the potential support these can provide to students' learning processes, in particular those of students classified as having MLD. We will also analyze some interactions that actually occurred between students and the software, highlighting important qualitative results from recent studies in which we have been involved.

1 Introduction

Since we will be discussing software with respect to students "with mathematical learning difficulties (MLD)" it is necessary to first explain how *unclear* the situation actually is around the issue of low achievement in mathematics and MLD. This will be done in the first section of this chapter, immediately followed by our opinion on ways in which software can address specific MLD. The rest of the chapter is divided into three other sections: one in which we describe the theoretical background we will be using to analyze the proposed examples of software; one in

E. Robotti (✉)
Università Di Torino, Torino, Italy
e-mail: elisabetta.robotti@unito.it

A. Baccaglini-Frank
Università Di Pisa, Pisa, Italy
e-mail: anna.baccaglinifrank@unipi.it

E. Faggiano et al. (eds.), *Innovation and Technology Enhancing Mathematics Education*, Mathematics Education in the Digital Era 9,
DOI 10.1007/978-3-319-61488-5_5

which we explore the design of specific digital environments to which we have contributed; and a last section presenting selected results from studies we have conducted with students using the previously analyzed software.

1.1 The Murky Notion of "Students with MLD"

When exploring persistent low achievement in mathematics from a cognitive point of view, most of the literature from the field of psychology investigates typical development of basic number processing, introducing terms for describing atypical situations. Terms used to refer to students in such situations include "developmental dyscalculia", "mathematical learning disability (or disorder)", among many others (e.g., Butterworth, 2005; Passolunghi & Siegel, 2004; Piazza et al., 2010). The definitions of these terms are still a topic of debate (e.g., Mazzocco, 2008), and the ways in which they are used in different studies is inconsistent. For example, Mazzocco and Räsänen (2013) note that "math learning disability (MLD) has been used as synonymous with DD [Developmental Dyscalculia] [...], but also as distinct from DD when MLD is used to refer to the larger category of mathematics difficulties (MD)" (ibid., p. 66). Even the use of the acronym MLD is not consistent, in that the "D" in some cases stands for "disabilities" and in others for "difficulties" (ML stands for "Mathematical Learning" in all cases). We attribute this, at least in part, to a problem described by Heyd-Metzuyanim (2013), according to which the "learning disability" construct does not afford to differentiate between difficulties that signal a stable disability in mathematics and those that are a result of inadequate teaching experiences or lack of sufficient exposure (also see González & Espínel, 1999; Mazzocco & Myers, 2003).

The bulk of studies conducted within the field of cognitive psychology use tests of different cognitive abilities (either cognitive domain specific or general) and investigate how scores derived from those tests correlate with students mathematical performance on standardized achievement tests (e.g., Geary, 1994, 2004; Nunez & Lakoff, 2005; Piazza et al., 2010; Andersson & Östergren, 2012; Szucs et al., 2013; Bartelet, Ansari, Vaessen, & Blomert, 2014). In this scenario it is not surprising that the cut off scores for diagnosing MLD vary from the 3rd to the 32nd percentile (Mussoli, 2009), and prevalence is reported between 1.3 and 13.8% of the population (see, for example, Kaufmann et al., 2013; Mazzocco & Räsänen, 2013; Watson & Gable, 2013).

It is beyond the scope of this chapter to delve deeper into these issues; for our purposes it suffices to consider students "having MLD" as students with persistent low achievement in mathematics (this is what the "D" in the acronym MLD will refer to in this chapter), who are at risk of being labelled by clinicians as "having a learning disability" or who have been diagnosed clinically with such a condition.[1] So any of these conditions are what we imply when using the acronym MLD in this chapter.

In Italy the percentage of these students diagnosed with learning disabilities is estimated between 3 and 5% (MIUR, 2011a) and over the last few years the percentages have been persistently increasing (MIUR, 2011b). Because of this phenomenon and because in Italy classrooms are completely inclusive,[2] it has become a more and more pressing issue to study and develop didactical practices appropriate for *all* students (Ianes, 2006; Ianes & Demo, 2013). Though frameworks for analyzing students' difficulties and/or for designing inclusive activities are still quite fragmentary, the literature rather consistently suggests that technology can support the learning of students with different learning characteristics (Edyburn, 2005; Baccaglini-Frank & Robotti, 2013; Robotti, Antonini, & Baccaglini-Frank, 2015), also in *inclusive* teaching settings, such as the Italian classrooms (Robotti & Ferrando, 2013).

1.2 How Can Software "Address" Specific MLD?

We must ask ourselves what it means to "address" students' learning difficulties. Once we will have agreed upon a meaning for this, we will be able to discuss how software can do it.

The paradigm used (at least in Italy) in special needs education, as has recently been argued by the Santi and Baccaglini-Frank (2015), is such that the teaching activity strives to allow the "special needs" student to reach as much as possible, according to his/her possibilities, the same objectives of "normal" students, thereby disregarding his/her identity and being "special" from many points of view (cognitive, social, communicative, emotional, perceptive…). The stand point behind this approach is that thinking and learning is purely in the functioning of the mind (or, according to neurosciences, in the brain) and that a deficit provokes a dysfunction that has to be recovered resorting to a variety of supports: technological, didactical, psychological and social. This leads to a homogeneization of all students' contributions, that tends to not take into account or value in any way alternative insight brought to the classroom by the special needs student. To overcome such approach, the authors proposed a paradigm shift: "Educational activity should aim at fostering a mode of existence in mathematics, i.e., being and becoming with others to make sense of the world also through mathematics. The aim of education should be to

[1]There are four types of learning disabilities recognized at the moment in Italy: dyslexia, dyscalculia, dysgraphia, dysorthographia (LEGGE 8 ottobre 2010, n. 170, Nuove norme in materia di disturbi specifici di apprendimento in ambito scolastico).

[2]In some "extreme" cases Italy grants a special education teacher to the student in need, who will sit next to the student during given hours of the student's regular school schedule.

allow all students to make sense of the world in spite of their particular conditions" (ibid., p. 222).

The described approach and the proposed paradigm shift are useful to keep in mind when considering two main directions in which the development of technological tools for MLD students seems to be going (at least in Italy): some software aim at strengthening a particular cognitive or mathematical ability, through repetitive tasks, designed for a one-to-one student-computer interaction, in an environment with constrained types of input and feedback—we will refer to this approach as for "rehabilitation"; while other software are designed to propose fundamental mathematical content (e.g., the notion of "variable" or "function") in ways that take advantage of particular hardware and software affordances. We will refer to this approach as "radical", because didactical material developed within it may propose (although they do not have to necessarily), more or less explicitly, radical changes in the mathematical curriculum and/or in the modalities in which certain content is proposed. Interactions with software designed according to the "radical" approach are frequently less constrained: tasks within the environment need to be designed by an educator (as they might not be part of the software), input and feedback may be given in various ways, and the role of the teacher becomes fundamental in mediating the meanings developed by the students within the environment.

Neither the "rehabilitation" nor the "radical" approach are necessarily one "better" than the other—of course to make any judgment of this sort we would have to make explicit the criteria according to which we are making such judgment—and both could be useful in supporting the learning of students with MLD. However, if our aim is to provide means for as many students as possible to make sense of the world, through mathematics, in spite of their particular conditions, it is inevitable to embrace, at least some of the time, the latter approach, when teaching. This approach is somewhat innovative in education, at least in the Italian panorama.

Since researchers in psychology and neuroscience have been designing, conducting and publishing research with rehabilitation software (e.g., Wilson, Revkin, Cohen, Cohen, & Dehaene, 2006; Wilson, Dehaene, Pinel, Revkin, Cohen, & Cohen, 2006; Butterworth & Laurillard, 2010), in this chapter we would like to focus mostly on software developed within the radical approach, which is innovative because it characterizes not only software design but also a general line of research regarding the development of didactical material that seems to be appropriate for inclusive mathematics education (see, for example Baccaglini-Frank & Poli, 2015a, b; Robotti, 2017). Software designed and adopted within the radical approach can also offer the student with MLD specific compensatory tools embedded within it, to alleviate the cognitive load of particular tasks in order for the student to be able to devote as many resources as possible to fundamental mathematical reasoning involved in the activity. However, these environments are not designed *only to compensate* certain cognitive difficulties. Within a software designed according to the radical approach there may exist sub-environments in which, through repetitive exercises, a specific ability or set of abilities may be strengthened. On the other hand, software developed primarily to strengthen a specific ability through repetitive exercise can be more difficult to use for fostering the development mathematical content within the radical approach. This is the case also because the closed, and in many cases fast, interaction between student and software does not leave much space for teacher-guided interventions.

In general there is no clear boundary between software designed according to either approach: we prefer to think of a spectrum with "radical" and "rehabilitation" designs at the extremes. Most software we can think of would be situated along the spectrum, more towards one or the other extreme. Moreover, there are significant variables, such as how the software is actually used or what role the teacher decides to play, that can contribute to shifting the software's placement within the spectrum, in either direction. In this sense, it can be possible to also use rehabilitation software within the innovative approach to special education presented above.

The perspective we are taking on how software can address specific MLD provides our rationale for analyzing how digital resources can support students in learning mathematics. The analyses will be carried out using a composite framework emerging from the notions of "Universal Design for Learning" and theories on channels for accessing and producing mathematical information.

2 Theoretical Background

In the field of mathematics education a number of frameworks have been developed, on one hand, to explain phenomena like "students experiencing learning difficulties in mathematics" from different perspectives, and others have provided tools for analyzing teaching-learning activity within technological settings (e.g., Lagrange, Artigue, Laborde, & Trouche, 2003; Noss & Hoyles, 1996; Bartolini Bussi & Mariotti, 2008). However, these theoretical tools are still quite fragmentary and very few have been adequately adapted and/or integrated to take into account findings (both practical and theoretical) from neighbouring fields such as cognitive psychology and neuroscience that have also been very active in investigating such phenomena. Notable exceptions are studies by the Unit of Instructional Psychology and Technology in Leuven, directed by Lieven Verschaffel (e.g., Vamvakoussi, Dooren, & Verschaffel, 2013); studies by Mulligan and her team based in Australia (e.g., Mulligan & Mitchelmore, 2013); and the work of the Center for Applied Special Technology (CAST), elaborating on the concept of Universal Design for Learning (Edyburn, 2005), which we will present in Sect. 2.2. Also, recent work of Karagiannakis and his colleagues contributes to establishing common grounds, at a cognitive level, attempting to transpose relevant aspects of the cognitive psychology literature into the field of mathematics education (Karagiannakis, Baccaglini-Frank, & Papadatos, 2014; Karagiannakis & Baccaglini-Frank, 2014; Karagiannakis, Baccaglini-Frank, & Roussos, 2017).

In particular in the Italian context, we have been active in trying to elaborate theoretical grounding for research on MLD students when teaching and learning include physical and digital artifacts (e.g., Baccaglini-Frank & Robotti, 2013; Baccaglini-Frank & Scorza, 2013; Robotti & Ferrando, 2013; Baccaglini-Frank,

Antonini, Robotti, & Santi, 2014; Robotti et al., 2015; Santi & Baccaglini-Frank, 2015; Robotti et al., 2015; Baccaglini-Frank & Bartolini Bussi, 2016). In the two following sections we will review some notions from the theoretical background of cognitive psychology that will be useful for the analyses in this chapter (Sect. 2.1), and review some principles and guidelines from the framework elaborated by CAST that will also be insightful in the analyses proposed in the rest of the chapter (Sect. 2.2). The relationship between these different frameworks will allow us to analyse how and why the use of technology can foster mathematical learning in all students who present MLD.

2.1 Means of Information Access and Production, with Particular Attention to Mathematical Information

Research in cognitive psychology has identified four basic channels of access to and production of information: the visual-verbal channel (verbal written code), the visual non-verbal channel (visual-spatial code), the auditory channel (verbal oral code), and the kinaesthetic-tactile channel (Mariani, 1996).

Italian research has indicated that most students with specific learning difficulties (or disabilities), not only in mathematics, encounter greatest difficulties in using the visual-verbal channel, especially those with dyslexia, and this conditions their development for preferring different channels (Stella & Grandi, 2011).

The importance of these different channels to access and produce information shifts the focus from simply "being able or not" to solve a certain task, to different paths and strategies adopted by the individual (whether successful or not) for approaching the task. This allows to explain mathematical difficulties not only in terms of "lacking abilities" but also in terms of necessity to use certain preferred modalities that lead the student to access, elaborate and/or produce information in a certain way.

Moreover, various studies in cognitive science point to a correlation between mathematical achievement, working memory (Raghubar, Barnes, & Hecht, 2010; Mammarella, Lucangeli, & Cornoldi, 2010; Mammarella, Giofrè, Ferrara, & Cornoldi, 2013; Szucs, Devine, Soltesz, Nobes, & Gabriel, 2013), and non verbal intelligence (DeThorne & Schaefer, 2004; Szucs et al., 2013). These findings suggest that non-verbal intelligence may partially depend on spatial skills (Rourke & Conway, 1997) and these can potentially be important in mathematical achievement, where explicit or implicit visualization is required.

We have found other theoretical stances advanced in mathematics education that are in line with the idea that means of access to and production of information, different from the visual-verbal one, can be very important in learning. In particular, these have pointed to the importance of experiences of a sensorial, perceptive, tactile and kinaesthetic nature for the formation of mathematical concepts (Arzarello, 2006; Gallese & Lakoff, 2005; Radford, 2003; see also Chap. The Coordinated Movements of a Learning Assemblage: Secondary School Students

Exploring Wii Graphing Technology). For example, Arzarello (2006), quoting Nemirovsky, points to how recent research in math education suggests that the paradigm of multimodality implies that "the understanding of a mathematical concept rather than having a definitional essence, spans diverse perceptuomotor activities, which become more or less active depending of the context" (Nemirovsky, 2003, p. 108). Also Radford (2003, 2006) highlights that the understanding of relationships between bodily actions carried out through artifacts (objects, technological tools, etc.) and linguistic and symbolic activity is essential in order to understand human cognition and mathematical thinking in particular.

A new framework for teaching and learning in the context of "special needs" has been developed, taking into account many of the perspectives advanced above, and suggesting that technology can facilitate all students' learning. The framework is built around the concept of *Universal Design for Learning*.

2.2 *Universal Design for Learning*

The Center for Applied Special Technology (CAST) has developed a comprehensive framework around the concept of Universal Design for Learning (UDL), with the aim of focusing research, development, and educational practice on understanding diversity and applying technology to facilitate learning (Edyburn, 2005). UDL includes a set of Principles, articulated in *Guidelines and Checkpoints*[3] that arise from CAST's review of current studies on how to reduce barriers in learning and to increase access to curriculum for all the students, including those with disability, giving all individuals equal opportunities to learn. The research grounding UDL's framework is that "learners are highly variable in their response to instruction. […] individual differences are not only evident in the results; they are prominent. However, these individual differences are usually treated as sources of annoying error variance as distractions from the more important "main effects"".[4] In contrast, UDL treats these individual differences as an equally important focus of attention. The UDL framework considers these findings to be fundamental to understanding and designing effective instruction.

As a matter of fact, "individuals bring a huge variety of skills, needs, and interests to learning. Neuroscience reveals that these differences are as varied and unique as our DNA or fingerprints. Three primary brain networks come into play:"[5] Recognition Networks, which refer to recognition tasks such as: How we gather facts and categorize what we see, hear, and read, Identifying letters, words; Strategic Networks, which refer to strategic tasks such as solve a math problem;

[3]For a complete list of the principles, guidelines and checkpoints and a more extensive description of CAST's activities, visit http://www.udlcenter.org.

[4]See http://www.udlcenter.org/aboutudl/udlevidence.

[5]See http://www.udlcenter.org/aboutudl/whatisudl.

Affective Networks, which refer to the affective dimension: How learners get engaged and stay motivated, How they are challenged, excited, or interested.

Linked to each of these brain networks, UDL advances three foundational Principles[6]: (1) provide multiple means of representation, (2) provide multiple means of action and expression, (3) provide multiple means of engagement. In particular, guidelines within the first principle have to do with means of perception involved in receiving certain information, and of "comprehension" of the information received. Instead, the guidelines within the second principle take into account the elaboration of information/ideas and their expression. Finally, the guidelines within the third principle deal with the domain of "affect" and "motivation", also essential in any educational activity. For our analyses in this chapter we will focus in particular on specific guidelines within the three Principles.[7]

Guidelines and checkpoints within Principle 1 (provide multiple means of representation), suggest proposing different options for perception and offering support for decoding mathematical notation and symbols (checkpoints 1.2, 1.3, 2.3). We will give examples of how this can be realized through different software. Moreover, guidelines suggest the importance of providing options for comprehension highlighting patterns, critical features, big ideas, and relationships among mathematical notions (checkpoint 3.2). We will identify various of such options in the remainder of the chapter. Finally, our analyses will give examples of how software can guide information processing, visualization, and manipulation, in order to maximize transfer and generalization (checkpoints 3.3 and 3.4).

Moreover, our analyses will provide examples of how guidelines from Principle 2 (provide multiple means of action and expression) can be incorporated into technology-based mathematical learning, in particular how different options for expression and communication supporting planning and strategy development can be offered (checkpoints 4.2 and 6.2). Finally, our analyses will show how certain software can recruit students' interest, optimizing individual choice and autonomy, and minimizing threats and distractions (checkpoints 7.1 and 7.3).

In the two following sections we will analyze specific examples of software, classifying them by the type of mathematical learning they are designed to address. The analyses highlight which kinds of compensatory tools each software offers the student and which kind of tasks could be designed in order for the student to be able to devote as many resources as possible to fundamental mathematical reasoning involved in the activity.

Each software will be introduced by a section looking into research around the particular way of thinking or concept or tool being targeted. The rationale for choosing the software presented is that each one was used by one of the authors in studies carried out in the context of special needs or inclusive mathematics

[6]For further details see: http://www.udlcenter.org/aboutudl/whatisudl/3principles.

[7]The items are taken from the interactive list at http://www.udlcenter.org/research/researchevidence.

education, focus of this chapter. In some cases one of the authors was also directly involved in the software design process, while in other cases a particular software was chosen among other existing ones because of its fit with the UDL principles.

3 Examples of Digital Environments to Promote the Development of Number Sense and Spatial Orientation

In this session we analyse different software promoting number sense and we report on results from a case study on learning special orientation by interacting with the software *Mak-Trace*.

3.1 Software to Promote "Number Sense"

According to various studies a characterizing feature of students with MLD is a lack of "number sense". Although there is no monolithic interpretation of *number sense* across the communities of cognitive scientists and of mathematics educators, and not even within the community of mathematics educators alone (e.g., Berch, 2005), there seems to be a certain consensus about its importance in mathematics education. Indeed the development of number-sense is seen as a necessary condition for learning formal arithmetic at the early elementary level (e.g., Griffin, Case, & Siegler, 1994; Verschaffel & De Corte, 1996) and it is critical to early algebraic reasoning, particularly in relation to perceiving the "structure" of number (Mulligan & Mitchelmore, 2013). Some crucial aspects upon which number sense is seen to rely, are: recognition of part-whole relationships, appropriate uses of fingers, and the development of a mental number line. We will describe these and explain how they can be promoted through software applications.

Part-whole relationships arise from what Resnick et al. (1991) have described as protoquantitative part-whole schemas that "organize children's knowledge about the ways in which material around them comes apart and goes together" (ibid., p. 32). The interiorization of the part-whole relation between quantities entails understanding of addition and subtraction as dialectically interrelated actions that arise from such relation (Schmittau, 2011), and recognizing that numbers are abstract units that can be partitioned and then recombined in different ways to facilitate numerical (also mental) calculation.

Literature from the fields of neuroscience, developmental psychology, and mathematics education indicate that using fingers for counting and representing numbers (Brissiaud, 1992), but also for accomplishing tasks that have no apparent connection to mathematics (Butterworth, 2005; Gracia-Bafalluy & Noel, 2008), can have a positive effect on the development of numerical abilities and of number-sense. The importance of the role attributed to the use of fingers in the development of number-sense by the quoted literature is highly resonant with the frame of embodied cognition, mentioned in Sect. 2.1. For example, hands and fingers can be used to foster development of the part-whole relation, in particular with respect to 5 and 10, in a naturally embodied way.

Development of a mental number line

Number sense has also been put in relationship with the development of an internal representation of the number line. A number of studies have explored a relationship between space and the processing of numbers (e.g., Pinel et al., 2004; Seron et al., 1992), suggesting that the (mental) number line model corresponds to an intuitive representation and to a natural translation of the sequence of (natural) numbers into a spatial dimension. This model can be used in more abstract (and potentially more general) processes compared to that of counting existing sets of objects, because, for example, it opens to the possibility of counting *any* number of objects and *any* object. The number line model is not a static representation, nor is it necessarily innate,[8] instead studies suggest that it evolves as the subject develops cognitively, and such evolution depends on cultural influences (see, for example, Zorzi, Priftis, & Umiltà, 2002).

Moreover, studies suggest that a solid mental representation of the number line provides students with a rapid and successful means of access to numerical information necessary for the development of a variety of arithmetical skills. The number line can also be an appropriate tool not only for calculation (mostly addition and subtraction) with numbers within 10 (which can also be done using hands and fingers) but also for dealing with numbers beyond 10, when hands and fingers no longer are sufficient.[9]

Finally, the number line is not made up of only natural numbers, but also all other real numbers, which include, for instance, fractions. However, frequently the position of numbers on the number line can become a cognitive obstacle: for example, placing fractions on the number line (mathematically this involves ordinal properties and the density in the field of rational numbers) is notoriously a difficult task for many students (Robotti et al., 2015).

Given these considerations on fundamental aspects that have been identified as promoters of number sense, we can assume that software designed to promote these aspects, may be used in one of two ways: to help prevent the emergence of MLD in young students (younger than 8), or to strengthen weaker "number sense" abilities of older students who have developed MLD. In the sections below we will describe two innovative examples of these kinds of software.

---

[8]For a more complete discussion see volume 42(4) of the *Journal of Cross-Cultural Psychology*.

[9]Sometimes fingers are used also to represent numbers larger than 10, but in this case the meanings referred to by different fingers must be different (for example 4 and 13 might be represented raising the same fingers: 1 on one hand and 3 on the other) which can be confusing for children.

3.1.1 Software Promoting Number Sense Through Fingers

Technology offers the possibility of embedding a number of features into software that can be significant in promoting number sense through the use of fingers. For example, thanks to touch and multi-touch screens, input may be given in terms of a number of fingers placed simultaneously on the screen, as a number of sequential taps (possibly on items in the stimulus), or as particular gestures (swipe, pinch, lasso/capture, …). Here we give an example of software that exploits such innovative potential.

TouchCounts,[10] an application for the iPad, is made up of two environments (Sinclair & Pimm, 2014; Sinclair & Zaskis, in press; see also Chap. Returning to Ordinality in Early Number Sense: Neurological, Technological and Pedagogical Considerations). Here we will briefly analyze the "Operating world" with respect to its design and potential of fostering development of number sense through fingers. In this environment the student can create autonomous numbered sets, here referred to as herds, by placing one or several fingers on the screen. This immediately creates a large disc encompassing all the fingers and including, in the middle, a numeral corresponding to the total number of fingers touching the screen. At the same time, every one of the fingers on the screen creates its own much smaller (and unnumbered) disc, centred on each fingertip. When the fingers are lifted off the screen, the numeral is spoken aloud and the smaller discs are then lassoed into a herd and arranged regularly around the inner circumference of the big disc. This design offers four representations (UDL Principle 1) of a number: visual non verbal (or analogical), symbolic (the numeral in the herd), auditory, and of course gestural (the number is represented by the number of fingers placed on the screen simultaneously). Moreover, the student is guided to perceive the herd a single entity made up of units through the movement of the small discs all together in either a clockwise or counter-clockwise direction.

The software also offers multiple means of action and expression (UDL Principle 2) because the student can act on the herds in different ways. For example, s/he can interactively drag herds, either to move them around on the screen or to operate upon them. After two or more herds have been produced they can either be pinched together (a metaphor for addition) or 'unpinched' (metaphor for subtraction or partition). When herds are pinched together they then become one herd that contains the small discs from each previous herd. The new herd is labelled with the associated numeral of the sum, which *TouchCounts* announces aloud. Moreover, the new herd keeps differentiated colors for the small discs coming from the previous herds. Similarly, the student can do an inverse pinch gesture to decompose a given herd into two herds. The gesture supports the idea of partitioning, or 'taking out' or 'removing', which, in turn, supports the idea of subtracting. The further the swipe travels, the more will be taken out from the starting herd. When the swiping finger is lifted, two new herds are formed and *TouchCounts* announces the number that has been taken out.

---

[10]See https://itunes.apple.com/us/app/touchcounts/id897302197?mt=8.

Students can engage with this software through different means (UDL Principle 3) —using gestures, through listening, visually—as they freely explore or approach a variety of tasks proposed by a nearby educator (e.g., "make *n* all at once", "count by *n*", "make the herds equal", "how many different ways can you make *n*?"). Indeed, the environment allows proposing many different types of tasks that can foster the development of number sense in different ways, through a "radical" approach.

3.1.2 Software Promoting Number Sense Through the Number Line

There are many software applications that propose representations of the number line: some are discrete containing only natural numbers, others continuous with marks such as those on the ruler, some are static and designed only for responding to specific tasks implemented within the application, while others are dynamic and allow various user interactions.

A first example we would like to analyze is *Motion Math*: *Fractions*,[11] an application for tablets. At the moment it is designed only for promoting processes involved in the estimation of fractions, exploiting both epistemological and cognitive analyses of fractions (Riconscente, 2013), emphasizing, on the one hand, the importance of using the number line to give coherence to the study of fractions and of whole numbers and, on the other hand, the neurological evidence of the mental number line discussed above (Zorzi et al., 2002).

Within this environment a number line appears on what looks like the "ground" together with a ball that can bounce (completely elastically) and that can be controlled by the gravity accelerator of the tablet that is, it responds to physically tilting the tablet, as if the ball had a weight. A fraction appears within the ball, which needs to be placed correctly on the line. The fraction is presented in different representational formats: it may be in the form *n*/*m*, or a decimal number, a percentage, or a shaded section of a circle. Successive hints are given if the user makes mistakes in positioning the fraction on the line. The app is designed as a game (the user gets points, passes levels, and "dies" when a mistake has been made even after all the hints). The ball's regular bounces constrain the user's response time, forcing each placement choice to be planned and executed in pre-determined and regular time intervals.

The application appears to be in line with a number of the UDL principles outlined in Sect. 2.2: multiple means of representation are provided and integrated (fractions are presented in different forms: as "*n*/*m*", as decimal numbers, as percentages, as parts of a whole, and as numbers on the number line), support is offered in the form of successive hints for finding the position of the given fraction on the number line, the successive hints highlight critical features of the relationship between the given representations of fractions and their position on the number line, no verbal skills are necessary because the channels activated for input and output of

[11]See the app *Motion Math: Fractions* at http://motionmathgames.com/motion-math-game/.

information are visual and kinaesthetic, distractions are minimized by the need to plan and give successive input according to pre-determined and regular time intervals.

Moreover, *Motion Math: Fractions* can be seen to exploit embodied learning and, in particular, the integrated perceptual-motor approach (Nemirovsky et al., 2012) in the development of the mental number line.

It is possible (and, we believe, advisable in many cases) to complement a student-software interaction with verbal guidance and successive discussion of each playing session. For example, in the episode presented in Bartolini, Baccaglini-Frank, and Ramploud (2014) the student, who had been diagnosed with various learning difficulties, including severe dyscalculia, was significantly helped by the introduction of a different way of reading the fractions in the falling ball. The teacher (second author) suddenly exclaimed: "Let's name the fractions as Chinese do!…[1/2 *falls*] Of two parts, take one!…[3/4 *falls*] Four parts, three!" and the student improved his performance very quickly, especially on unitary fractions (e.g., 1/5). Similar episodes have since been observed with other low achievers.

In this example we can observe that providing options for mathematical expressions and symbols by language and different linguistic expression, can be effective for overcoming some difficulties in math comprehension (according to Principles 1 and 2 of UDL framework). We note that in the case described above the verbal expression that identifies the fraction expresses at the same time a process for constructing (and thus placing) the fraction that follows a same order.

3.2 Spatial Orientation and Non-verbal LD

A possible source of difficulties in mathematical learning is what has been referred to as a non-verbal (or visual-spatial) LD (e.g., Mammarella et al., 2010; Andersson & Östergren, 2012; Mammarella et al., 2013). An ability that may be weaker in these students is perspective-taking (Piaget & Inhelder, 1967; Clements 1999), that is being able to embrace different frames of reference based on one's self or on external points of reference, is fundamental both in everyday life and in instruction. The importance of such ability is declared, for example, in the Italian National Curriculum Indications (MIUR, 2011a, b) relative to mathematical learning about *Space and Figures*. Developing the perspective-taking ability may not be straightforward: it involves a transition from "perceptual space" to "representational space" (Piaget & Inhelder, 1967), as well as "connecting different viewpoints" (Clements, 1999, p. 3).

While children showing typical development seem to have acquired such ability by the end of primary school, in some children with MLD—including developmental dyscalculia (e.g., Mazzocco & Räsänen, 2013)—the development of perspective-taking, among other abilities, may be delayed and/or deficient.

Software environments that seem particularly appropriate for addressing perspective taking are microworlds, such as *Logo* (Papert, 1980). The potential of

Logo-like microworlds for fostering learning in students with persistent difficulties in mathematics is documented in the literature. In particular, Vasu and Tyler found that Logo may foster the development of spatial abilities and of critical thinking skills (Vasu & Tyler, 1997), and various other researchers have reported several potential benefits of using Logo with students who have learning difficulties (Atkinson, 1984; Maddux, 1984; Michayluk & Saklofske, 1988; Russell, 1986), especially using a more structured, mediated approach (Ratcliff & Anderson, 2011).

Below we describe design features of a Logo-like microworld, Mak-Trace, an environment we used to analyze cognitive processes involved in juggling different frames of reference of students with non-verbal difficulties.

3.2.1 The Logo-like Microworld Mak-Trace

Mak-Trace is an environment in which a character can be programmed to move and draw on a grid. The grid is 10×15 and the character can only be programmed to go forwards (F) or backwards (B) (of the distance of one side of a square of the grid at the time) or to turn 90° clockwise (R) or counterclockwise (L). The characters can be dragged on the grid with a finger to choose a starting position and then they will, by default, leave a trace mark as they move according to the commands in the programmed sequence (see Fig. 1). It is also possible to program the character so that it does not leave a trace mark on the grid, by inserting appropriate commands in the programmed sequence. The commands appear as icons that have to be dragged and placed on a vertical bar that represents the programmed sequence. This design proposes different representations (UDL Principle 1) corresponding to the

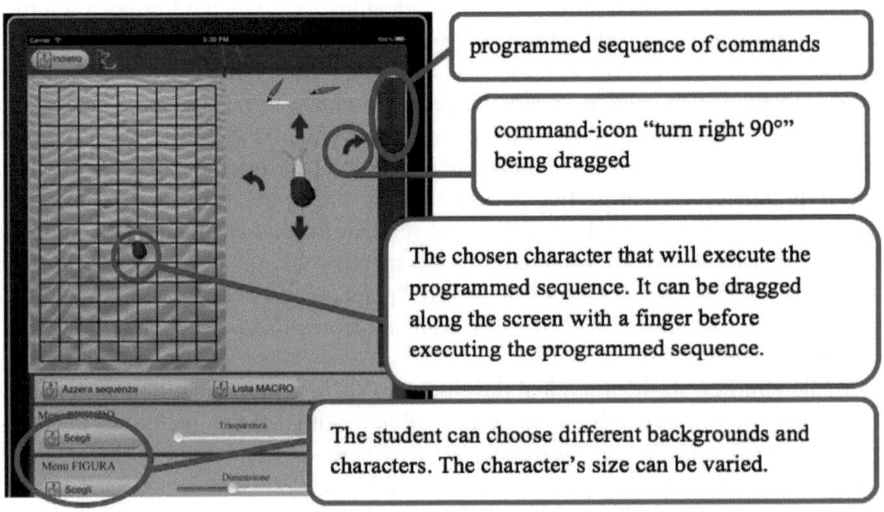

Fig. 1 Main screen in Mak-Trace, where the student can program his/her character

movements of the snail on the grid: a "draggable" arrow-symbol, a movement of the character on the grid, a segment (or point) traced on the grid.

An aim in designing Mak-Trace was to create an environment accessible to young children, or students with learning difficulties or disabilities, especially of a visual-spatial nature, by offering an intuitive iconic programming language. Students can act on the environment in different ways (UDL Principle 2): dragging the character on the grid with their finger, or dragging command icons to into a sequence to make a "program". Of course the student can also interact verbally with a nearby educator.

The fact that the command-icons can be treated as objects can make it natural to assign symbolic names to each of them in order to quickly describe a programmed sequence, orally or by writing on paper (Principle 2 of the UDL framework). This practice can be proposed and pursued by an educator using Mak-Trace with her students, and it may help students make use of a pre-algebraic language that can be quite useful in certain tasks involving generalization.

Another design choice is that Mak-Trace gives no feedback in terms of movements of the character until the student touches "GO". At this point the character executes the whole list of commands in the constructed sequence. To change the constructed sequence, the student has to go back to the "programming mode": automatically the character goes back to its original position and all trace marks are cleaned off the screen. This choice was made to foster planning and spatial orientation abilities. In particular, the student has to visualize what the character will do as she is programming, and where the character will be at each step of the programmed sequence, before actually executing the sequence. These design choices were made in accordance with the UDL Checkpoints 4.2 ("Optimize access to tools and assistive technologies") and 6.2 ("Support planning and strategy development").

In Mak-Trace the perspective-taking ability consists in embracing the character's moving frame of reference. To exemplify how working in this environment can be beneficial to students who experience difficulties in perspective-taking, we will revisit some critical episodes from a case study (Baccaglini-Frank et al., 2014; Santi & Baccaglini-Frank, 2015).

3.2.2 The Case of Filippo

Filippo was 15 years old and had been diagnosed by clinicians as having MLD including dyscalculia and severe dyslexia. From the accounts of his special education teacher, he also was not able to read maps or to give directions, however he did not show difficulties in recognizing or naming his left and right hands. He had a short attention span and little—if any—interest in the activities proposed during math class. Furthermore he suffered from very low self-esteem and sense of self-efficacy. We developed a protocol so that Filippo would work with Mak-Trace when he met with his special education teacher, for five weeks, either once or twice each week. The tasks were designed based on two hypotheses: we expected Filippo's perspective-taking ability to be weak at least initially, but all the same we expected that interacting with the software under supervision of the teacher could enhance his abilities to plan, visualize, and give directions, potentially through means different than his perspective-taking ability. Here we briefly report on the two tasks Filippo carried out: (1) describe the relationship between sequences of commands in Mak-Trace, and the movements and trace mark left by the snail; (2) program the snail to draw a square.

During the first task Filippo initially thought that the arrow commands "go forward", "go backward", "turn right", "turn left" (F, B, R, L) made the snail go forward, backward, right, and left, where these directions are relative to Filippo's front, back, left and right, or possibly to "absolute" directions, like north, south, east, west. Therefore Filippo was not able to construct a sequence of commands to make the snail draw a given path. For over half an hour he struggled to relate the brief sequences of commands he programs to their representation on the grid. He did not seem to be aware of any reference frames other than his own until the teacher intervened, in the interaction that follows.

Filippo: it went backwards, not upwards [...]
Teacher: so what do the little arrows refer to?
Filippo: it depends on how the snail is oriented.

This was a decisive moment which lay the foundations for Filippo's conception of the snail's perspective. However, Filippo still mostly relied on trial and error, embracing the snail's perspective as long as the snail is not oppositely oriented, which he was confronted with in the task of making the snail draw a square.

The first time Filippo tried to program the snail he was able to program the sequence correctly for the first two sides of the square, then he uses (incorrectly) the commands B and R, correct in his frame of reference, but not in the snail's; while the fourth side, horizontal in Filippo's frame, is programmed correctly. It is interesting that he used opposite commands for the first and third sides (F and B, respectively), while for the second and fourth he used the same command (F). The effect of this programmed sequence is shown in Fig. 2a.

The second time Filippo tried to program the sequence, he composed: FFFFLFFFFL [hesitated, inserted L, erased it, and with the index of his right hand made the gesture of a counter clockwise turn] FFFF [he said: "I have to always keep the" and made another counter clockwise turn gesture with his right hand] RFFFF (Fig. 2b). The feedback from Mak-Trace (snail moving on the screen and leaving a mark on the grid) confirmed that three sides were now correctly programmed. However Filippo made a mistake again on the rotation when the snail is oppositely oriented. This behavior suggests that indeed Filippo had a weak perspective-taking ability.

However, our second hypothesis was also confirmed, as Filippo, on his own, interacting with the software, developed alternative strategies for managing the different frames of reference. A first strategy is developed to finally solve the square drawing task. This time Filippo re-wrote the sequence: FFFFLFFFF [he made the gesture of a counter clockwise turn with his right hand] LFFFF...[he rotated the

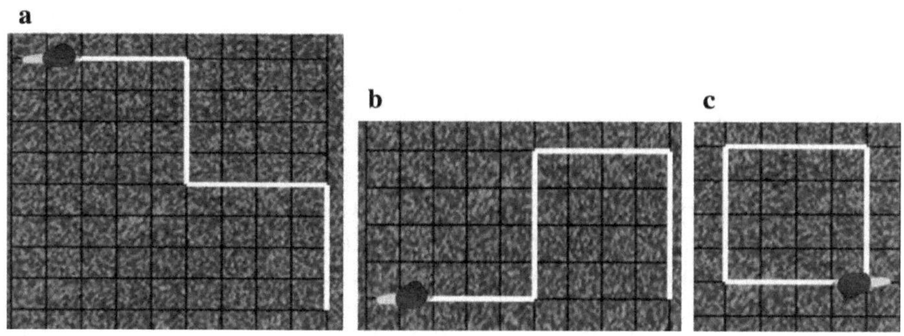

Fig. 2 Effect of Filippo's first, second and final programmed sequence

iPad so that his frame coincided with the snail's, observing the screen he rotated his right hand counter clockwise]. Then he completed the last turn and side.

> Filippo: Done, I found it […] no, I got…lost […] when it is turned around…it goes opposite [*clockwise* rotation gesture with the right hand] so…if I want it to go here [horizontal gesture from left to right with the left hand] … oh, I don't know, I'll try this [RFFFF]… no wait, because this otherwise is like before [he substitutes R with L].

The sequence was correct (Fig. 2c).

Rotating the iPad is a gesture that reveals how Filippo is now aware that he should consider the snail's frame of reference, and that this frame is oppositely oriented with respect to his (at the moment of the rotation). It is as if Filippo was aware of not being able to *feel* the snail's frame of reference when it is "too different" from his own (oppositely oriented), so he figured out a way of physically making the frame of reference of his body match the one of the snail. This allowed him to overcome his disorientation and to successfully complete the task.

4 An Example of Digital Environment Promoting Algebraic Abilities

We now briefly discuss learning difficulties in algebra. In this discussion algebra will be the chosen learning object (Principle 1 of UDL framework), and we analyse potentialities of the software *AlNuset*, showing how they played out during a case study. In this sense, according to Principles 1 and 2 of the UDL framework, we will analyse how *AlNuset* introduces both multiple means of representation and multiple means of actions and expression in order to help students grasp the meaning of some algebraic notions. The analysis will be focused, in particular on the MLD students' difficulties.

With a significant percentage of students, the current teaching of algebra seems not to be sufficient to effectively develop skills and knowledge to master this domain of knowledge (Sfard & Linchevski, 1992; Kieran, 2006). Here, we focus on the students' difficulties in algebra considering, in particular, students with MLD. These students can have severe difficulties in arithmetic, (Butterworth, 2005), however, there are also areas of mathematics, which do not depend so much on manipulating numbers, such as algebra, geometry and topology.

Indeed, some studies on MLD students have shown that there is dissociation between the recovery ability of arithmetic facts, which is compromised, and algebraic manipulations, which are intact (Hittmair-Delazer, Sailer, & Benke, 1995; Dehaene, 1997). Thus, there is evidence for the existence of two independent processing levels of mathematics: a formal-algebraic level and an arithmetic-numeric level (Dehaene, 1997). Moreover, neuroimaging results, focusing on the algebraic transformations, have highlighted how the visual-spatial areas of the brain are activated at the expense of those devoted to language. For example, it has been shown that when we solve equations, the expressions are manipulated mentally by means of a visual elaboration rather than through verbal means (Landy & Goldstone, 2010). Such neuroscientific results can help us analyze the difficulties of students with MLD in algebra.

Many students' difficulties in algebra, including difficulties in controlling algebraic manipulation (e.g., Robotti & Ferrando, 2013), seem to be due to a lack of grasp on the meaning of the notions involved (Arzarello, Bazzini, & Chiappini, 1994). Recent studies in math education have suggested that the construction of mathematical knowledge, as a cognitive activity, should be supported by the sensori-motor system activated in suitable contexts (Arzarello, 2006). Indeed, according to Nemirovsky (2003), the understanding of a mathematical concept spans diverse perceptuomotor activities, which become more or less active depending of the context. Thus, the construction of meaning can be seen as based on a rich interplay among three different types of semiotic sets: speech, gestures and written representations (Radford, 2003, 2006). Studies concerning both the algebraic domain (Chiappini, Robotti, & Trgalova, 2009; Chaachoua et al., 2012) and the geometrical domain (Goldenberg, Cuoco, & Mark, 1998) suggest using educational tools through which images can be constructed and managed (dynamically or statically), exploiting mainly visual non-verbal rather than (or together with) verbal means. This is in accordance with the UDL principle of providing multiple means of action and expression (Principle 2).

We will show how the software AlNuSet (Algebra of Numerical Sets) can be used to make algebraic notions explicit, and to construct their meanings dynamically, while involving all the students in a classroom, as much as possible (Baccaglini-Frank & Robotti, 2013). In particular we will look at how AlNuSet can be used in relation to the algebraic notions of variable, unknown, algebraic expression, equation and solution of an equation, and the formal solution of an equation can be addressed with the support of AlNuSet.

4.1 AlNuSet to Construct Algebraic Meanings: Examples to Inclusive Education

AlNuSet was designed for secondary school students (from age 12–13 to age 16–17) and it is made up of three separate environments that are tightly integrated: the Algebraic Line, the Algebraic Manipulator, and the Cartesian Plane. We will describe some features of these environments, with particular attention to the Algebraic Line and the Algebraic Manipulator, through examples of activities,[12] stressing their support for the conceptualization of algebraic notions in MLD students.

Variable and dependent expressions

On the Algebraic Line it is possible to place variables and expressions that depend from them. To do this, the user has to type a letter, for example, "x", and a mobile point will appear on the line. The point can vary within the chosen set of numbers (natural, whole, rational, or real[13]) and variation can be controlled directly by the user through dragging. This feature was designed so that important aspects of the notion of *variable* could become embodied. Moreover, it is possible to construct expressions on the line that depend on a chosen variable, for example, $2x + 1$. This dependent expression cannot be acted upon directly, but it will move as a consequence when x is dragged. The dependent expression will assume the positions on the line that correspond to the values it takes on when the dependent variable takes on the value it is dragged to (Fig. 3).

We note that the functionalities described propose different representations (UDL Principle 1) and they are designed to foster for the user a mediation of the algebraic concepts of *variable* and *dependent expression*, through a dynamic model that can be acted upon (UDL Principle 2). The mediation can occur thanks to visual and kinaesthetic channels, without the need of visual verbal means (written language). The construction of the concept realized as so may allow students, and especially students with MLD, to find mnemonic references that are appropriate for their cognitive style. This allows them to start using representations of the fundamental algebraic concepts at stake, and possibly to place and retrieve them from long term memory in a more effective way. AlNuSet allows to address "typical" topics in the secondary school algebra curriculum; in particular, in the following section we will analyze how *equations* can be addressed.

Equations

Let us consider a common task: "Solve the Eq. $3x-5 = 13$", or—stated in a possibly less common way—"Find the values of x for which the expression $3x-5$ is equal to 13".

---

[12]For a more detailed description of these environments see www.alnuset.com.

[13]Of course the representations of the numerical sets are accomplished on a computer, so the sets are actually finite and discrete, but they simulate—with some limitations—the properties of the number sets they represent.

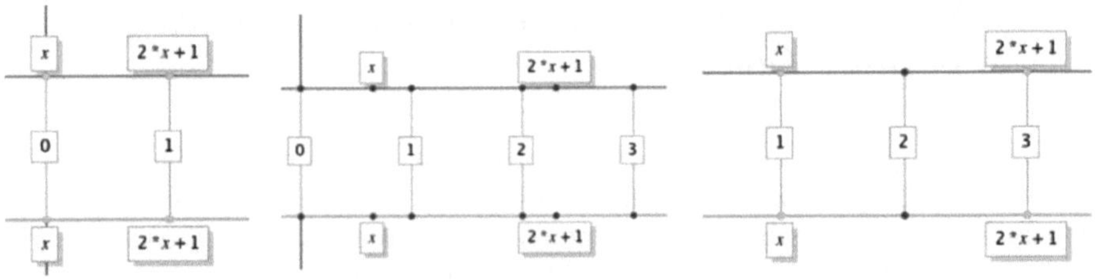

Fig. 3 The movement of the variable x on the Algebraic Line produces the movement of the dependent expression $2x + 1$ on the line

Solution on the Algebraic Line

Solving this equation on the Algebraic Line requires observing for which values of x the expression $3x{-}5$ (represented as a mobile point on the line) coincides with the number 13. When trying to verify the equality of expressions, dragging x is accomplished with a specific objective: that of trying to make the expressions coincide, that is, to make them take on a same value, becoming thus associated to a same "post-it" (yellow rectangle in Fig. 4). If the dragging is done with this objective, the variable assumes a meaning similar to that of *unknown*, that is of letter of which values need to be found in order to make the equality true. This allows students to act on the representations in different way, according to UDL Principle 2.

In Fig. 4 we can observe what happens on the Algebraic line as the point "x" is dragged.

The possibility of solving an equation through a perceptive kinaesthetic approach (dragging x along the line) without directly using a solution algorithm can help students concentrate their attention on the meaning of equation and its solutions. The Algebraic Line in AlNuSet was designed with this aim, which it attempts to reach through specific signs and functionalities embedded in it. Among these there is the possibility of dragging the point corresponding to "x", the visualization of "post it" markers containing values on the line and the constructed expressions that correspond to them (Fig. 4), the color of the dot corresponding to the equation (Fig. 4a and b). In particular this last feature is an example of how a visual non-verbal channel is used to give feedback to the student, guiding his/her construction of meaning of *solution of an equation.*

Features like the dot changing color and the yellow "post-it" signs, supporting the comprehension and the construction of meaning for algebraic notion and relationships involved, are examples of how AlNuSet's design seems to be well in line with the UDL principle advocating "multiple means of representation" (Principle 1). Indeed, they support perception providing the representations for algebraic notions through different modalities (e.g., through vision, dynamic image, touch…); and in a way that will allow for adjustability by the user (e.g., dragging the point corresponding to x as often as the user wants). Such multiple representations not only ensure that algebraic notion is accessible to MLD student, but also easier to comprehend for many others.

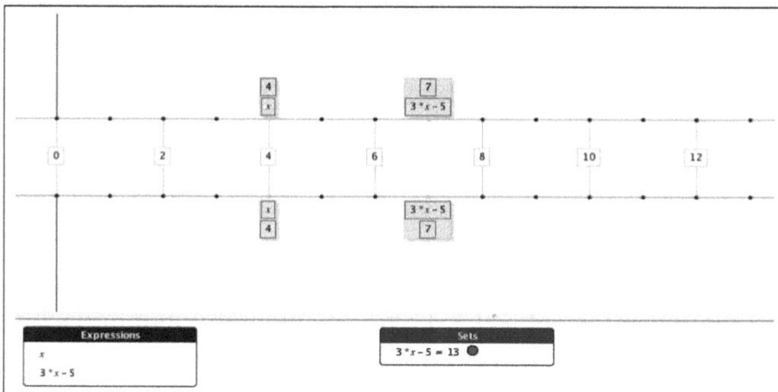

a Moving x along the line, the expression moves as a consequence. For example, when x is on 4, the expression 3x-5 is on 7. Notice that in the window "Sets" the equation is followed by a red dot that indicates that the value currently assumed by x does not make the equality true.

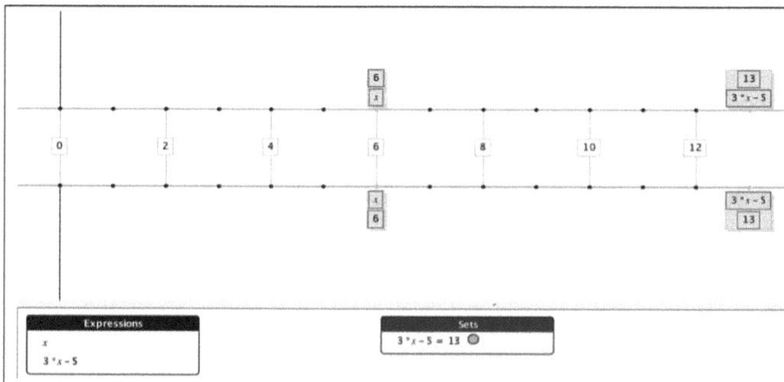

b When x in on 6, the equality is true because the expression 3x-5 is on 13. Notice that the dot associated to the equation in this case turns green.

Fig. 4 A way of solving the Eq. $3x{-}5 = 13$ on the Algebraic Line

Another functionality of AlNuSet that can be useful for the construction of a solid meaning of solution of an equation is the command "E = 0", in the environment Algebraic Line. This command allows the student to ask the system to calculate the roots of a polynomial (to read more about this functionality visit www.alnuset.com). This functionality can help the student tackle the "truth value of an equation", alleviating his/her cognitive resources from the burden of calculation procedures associated with the solution algorithms of an equation. This can be appreciated, for example, thinking about the cognitive load—excessive for some students—associated with the application of quadratic equations. Indeed, many students with MLD have trouble both with arithmetic calculations and with memorization and execution of procedures. The more complex a procedure is, greater are the difficulties for these students to retrieve the steps involved and to execute them. The "E = 0" functionality of AlNuSet allows these students to focus their attention on the cognitive task related to the meaning of solving an equation, in terms of searching for truth values of the equation, as opposed to dispersing their cognitive resources only on the calculation, loosing track of most (or all) meaning (Robotti, 2014; Robotti & Ferrando, 2013).

Given these features and ways in which their use can be integrated in approaching mathematical situations, the Algebraic Line can be used as a tool that can help lighten the cognitive burden involved retrieving and carrying out procedures, and allow the student to focus most of his/her cognitive efforts on the construction of the algebraic meanings at stake, favoring autonomy in approaching algebra. This is in agreement with UDL Principle 2 and, in particular, with the idea to provide option for comprehension: guiding information processing, visualization, and manipulation; maximizing transfer and generalization.

4.2 The Case of Eleonora

We now present the case of a student we will call Eleonora using the Algebraic Line of AlNuSet, carried out by the first author. She was 26 years old at the time of the study and had obtained her first diagnosis of dyscalculia the same year.

Before proposing the use of AlNuSet, one of the questions the interviewer asked Eleonora was the following: "*When 3 is added to 3 times a certain number, the sum is 28; find the number*".

Eleonora did not set up an equation, but proceeded by subtracting 3 from 28 (obtaining 25) and then dividing by 3, "undoing" the operations stated in the problem text. She then tried to prove the arithmetical equality (in Fig. 5) through "trial and error", approximating the value of $\frac{25}{3}$ to 8.333… She preferred to do this in spite of what she had been taught in various algebra classes where many examples of verbal texts of this type had been given and transformed into equations, such as $3x + 3 = 28$.

The researcher (first author) advanced the hypothesis that Eleonora had not developed a strong enough (if any) mathematical meaning of the notion of equation, possibly also due to the fact that she had trouble managing the typical procedures given to her during regular courses for solving first and second degree equations. The intervention proposed to Eleonora therefore was planned as a sequence of activities with the Algebraic Line in AlNuSet aimed at developing the mathematical

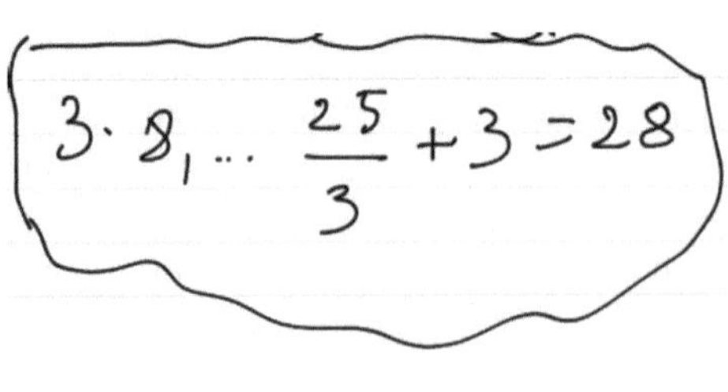

Fig. 5 Eleonora's attempt to solve the interviewer's question

meaning of equation and of solution of an equation. In the following excerpt we show Eleonora responding to the researcher's (R) question: "For which value of "a" is the expression $2 \times a$ equal to 8?".

1. E: Right now we can see that "a" changes value,… it changes value if I drag it
2. R: For which value of "a" is the expression equal to 8?
3. E: The expression is equal to 8… that is 2*a is equal to 8…
4. E: If I move it along the line, I am looking for the right value, where the letter matches
5. E: For example, I discovered that if I place "a" on 3…if I give "a" the value 3… 2*a is 6
6. E: Instead, if I put "a" on 4, 2*a is 8… because I'm multiplying […]
7. R: What did you get? [Referring to the colored dot associated to the equation in Sets window]
8. E: A verification. It's a check, if I drag "a", the red dot shows that I make a mistake
9. E: …if I drag "a", if I change the value of "a", the red dot shows that I make a mistake
10. E: Because, in this moment, 2*a equal to 8 is not true
11. E: There isn't an equality. Because I'm on 2*a equal to 10, if I give "a" the value 5

The solution to the problem is developed through a visual-spatial kinaesthetic approach in AlNuSet. Here, new representations (algebraic expressions, post-it, colored dots…) and different ways to act on them are provided, as proposed by UDL Principle 1 and Principle 2. As matter of fact, manipulating the expression $2 \times a$ on the line allows Eleonora to associate meaningful (to her) dynamic representations of the notions of *variable, unknown, equation and solution.*

Indeed we can observe that the verbal utterances used by Eleonora first refer to perceived aspects of the solution to the problem. Examples of such utterances are: "If I place "a" on 3…" (5) or "If I put "a" on 4…" (6). Later she seems to be attributing to "a" characteristics of an unknown: "if I give "a" the value 3…" (5), "in this moment, $2 \times a$ equal to 8 is not true" (10).

In intervention (6), we can also observe that Eleanor manages to relate the truth of the equation obtained by assigning to "a" the value of 4, with the arithmetic operation in $2 \times a$, which guided her first solution strategy (in the pre-testing phase). Thus, dragging "a" along the line until the value 4, she finds a link between the "meaning of an equation solution" with the "arithmetic procedure".

The construction of these meanings seemed to become more and more stable throughout the intervention, that is Eleonora was able to access and retrieve the meanings constructed within the Algebraic Line environment even months after the end of the intervention. This suggests a transfer to long term memory. Referring to the UDL principles, this environment seems to have successfully provided for Eleonora multiple means of representation, in this case offering dynamic representations of algebraic objects on the Algebraic Line of AlnuSet. Moreover, it provided multiple means of action and expression, exploiting the various functionalities through which Eleonora could act on receiving instantaneous feedback from the system. Making sense of such feedback Eleonora was able to give meaning to and manage the process of the solution of equations.

5 Conclusion

Specific theoretical frameworks in mathematics education research for the use of technology for fostering mathematical learning of students with MLD are still quite fragmentary. Moreover, very few have been integrated with findings from fields such as cognitive psychology and neuroscience, fields that have also been very active in investigating such phenomena. Therefore we felt the need to turn to more general theoretical notions related to different research fields. Among them, the idea of different means of information access and production, related to research in cognitive psychology, the three primary design principles of the Universal Design for Learning framework, which we refer to specific software's' design, and the paradigm of multimodality, related to research in math education, according to which experiences of a sensorial, perceptive, tactile and kinaesthetic nature are essential for the formation of mathematical concepts.

If we turn back and think about the analyses of students' interactions with selected software, we can again trace down our effort of seeking out evidence, within each particular mathematical learning context, of the usefulness of design choices, interpreted as aligned with the general UDL framework. In the case of Filippo, use of Mak-Trace, mediated by the teacher, helped the student develop personal strategies to solve problems concerning perspective-taking ability that initially he found unsormountable. These strategies later were endorsed also by his regular mathematics teacher. The analysis pointed to specific instances in which the software allowed the student to avoid the use of symbolic language and to rely on his sensorimotor activity in an interplay between movement, gestures and language (multiple means of action and expression—Principle 2). Moreover, similarly to what has been described for Logo, Mak-Trace appeared to be highly engaging (Principle 3), helping the student to "remain absorbed in a task for a period of time; … tolerate a period of confusion (with appropriate support);… use errors as a source of information about what to try next" (Russell, 1986, p. 103). In the case of Eleonora we highlighted how the environment seemed to successfully provide her with multiple means of representation (Principle 1) of algebraic objects on the Algebraic Line (for example, mobile points representing variable, expressions or unknowns, or the "yellow square" indicating expressions that refer to the same value/point on the line), and multiple means of action and expression (Principle 2), leading to instantaneous feedback from the system (for example, the movement induced by dragging a point on the line).

In general, we showed how the software applications analyzed provide multiple means of representation (Principle 1 of UDL framework), multiple means of action and expression (Principle 2) and multiple means of engagement (Principle 3), meeting specific checkpoints within each of these principles. To complete the analyses of each environment we also felt the need to add discussions of important literature on the learning of the specific mathematical content involved. This is because it is well known in mathematics education that the learning of different concepts or ways of thinking in mathematics can involve the activation of different cognitive processes in the students; and for the learning to be promoted effectively, it implies specific pedagogical content knowledge for teaching (Ball, Lubienski, & Mewborn, 2001) which the context of MLD includes information on cognitive issues involved in the learning of the specific mathematical content.

For this analysis, we referred mostly to software developed within the "radical" approach, according to which new ways of approaching specific mathematical content can also lead to changes in the organization of the mathematical curriculum or in the ways in which certain content is proposed (see, for instance, the notions of *variable* or *unknown* addressed in AlNuSet). As of today, we have only taken some initial steps towards reaching a framework to analyze the use of technology for fostering mathematical learning of students with MLD, and we definitely have yet a way to go in this direction. Until now we have (1) looked for ways of implementing checkpoints from the UDL principles designing software we collaborated to produce, and we have (2) looked for evidence of the usefulness of such design choices analyzing students' interactions with the software. These two tasks are still far from straightforward and necessitate a good deal of discussion and interpretation of the checkpoints of the UDL framework, because these are stated in very general terms. This of course makes them applicable to a number of different learning contexts (other than mathematics), but it costs their meaningfulness within the domain of mathematical learning, or even within more specific contexts, like learning natural numbers, learning about geometrical figures, or learning to solve quadratic equations.

We believe it is yet premature to propose a new coherent framework through which to look at technology mediated learning in the presence of MLD, but at the moment we see the intertwining of the different theoretical notions used for the analyses of the software and of students' interactions with the software as effective in giving insight into how and why some innovative software can foster mathematical learning for students with MLD.

References

Andersson, U., & Östergren, R. (2012). Number magnitude processing and basic cognitive functions in children with mathematical learning disabilities. *Learning and Individual Differences, 22,* 701–714.

Arzarello, F. (2006). *Semiosis as a multimodal process* (pp. 267–299). Numero Especial: Relime.

Arzarello, F., Bazzini, L., & Chiappini, G. P. (1994). Intensional semantics as a tool to analyse algebraic thinking. *Rendiconti del Seminario Matematico dell'Università di Torino, 52*(1), 105–125.

Atkinson, B. (1984). Learning disabled students and Logo. *Journal of Learning Disabilities, 17*(8), 500–501.

Baccaglini-Frank, A. (2015). Preventing low achievement in arithmetic through the didactical materials of the PerContare project. In X. Sun, B. Kaur, & J. Novotná (Eds.), *ICMI Study 23 Conference Proceedings* (pp. 169–176). Macau—China: University of Macau.

Baccaglini-Frank, A., & Bartolini Bussi, M. G. (2016). Buone pratiche didattiche per prevenire falsi positivi nelle diagnosi di discalculia: Il progetto PerContare. *Form@re, 15*(3), 170–184. doi:http://dx.doi.org/10.13128/formare-17182

Baccaglini-Frank, A., & Poli, F. (2015a). *Migliorare l'Apprendimento. Percorso per l'insegnamento in presenza di BES al primo biennio della scuola secondaria di secondo grado.* Novara: DeAgostini Scuola.

Baccaglini-Frank, A., & Poli, F. (2015b). *Migliorare l'Apprendimento. Percorso per l'insegnamento in presenza di BES al secondo triennio della scuola secondaria di secondo grado.* Novara: DeAgostini Scuola.

Baccaglini-Frank, A., & Robotti, E. (2013). Gestire gli Studenti con DSA in Classe Alcuni Elementi di un Quadro Comune. In C. Cateni, C. Fattori, R. Imperiale, B. Piochi, & P. Vighi (Eds.), *Quaderni GRIMeD n. 1* (75–86).

Baccaglini-Frank, A., & Scorza, M. (2013). Preventing learning difficulties in early arithmetic: The PerContare project. In T. Ramiro-Sànchez & M. P. Bermùdez (Eds.), *Libro de Actas I Congreso Internacional de Ciencias de la Educatiòn y des Desarrollo* (p. 341). Granada: Universidad de Granada.

Baccaglini-Frank, A., Antonini, S., Robotti, E., & Santi, G. (2014). Juggling reference frames in the microworld Mak-Trace: The case of a student with MLD. Research Report in C. Nicol, P. Liljedahl, S. Oesterle, & D. Allan (Eds.), *Proceedings of the Joint Meeting of PME 38 and PME-NA 36, 2* (81–88). Vancouver, Canada: PME.

Ball, D. L., Lubienski, S., & Mewborn, D. (2001). Research on teaching mathematics: The unsolved problem of teachers' mathe-matics knowledge. In V. Richardson (Ed.), *Handbook of research on teaching* (4th ed., pp. 433–456). New York: Macmillan.

Bartelet, D., Ansari, D., Vaessen, A., & Blomert, L. (2014). Research in developmental disabilities cognitive subtypes of mathematics learning difficulties in primary education. *Research in Developmental Disabilities, 35*(3), 657–670.

Bartolini, M. G., Baccaglini-Frank, A., & Ramploud, A. (2014). Intercultural dialogue and the geography and history of thought. *For the Learning of Mathematics, 34*(1), 31–33.

Bartolini Bussi, M. G., & Mariotti, M. A. (2008). Semiotic mediation in the mathematics classroom: Artifacts and signs after a Vygotskian perspective. In L. English et al. (Eds.), *Handbook of international research in mathematics education* (2nd ed., pp. 746–783). New York and London: Routledge.

Berch, D. B. (2005). Making sense of number sense: Implications for children with mathematical disabilities. *Journal of Learning Disabilities, 38*(4), 333–339.

Brissiaud, R. (1992). A toll for number construction: Finger symbol sets. In J. Bidaud, C. Meljac, & J.-P. Fischer (Eds.), *Pathways to number: Children's developing numerical abilities*. New Jersey: Lawrence Erlbaum Associates.

Butterworth, B. (2005). The development of arithmetical abilities. *Journal of Child Psychology and Psychiatry, 46,* 3–18.

Butterworth, B., & Laurillard, D. (2010). Low Numeracy and Dyscalculia: Identification and intervention. *ZDM Mathematics Education, 42,* 527–539.

Chiappini, G., Robotti, E., & Trgalova, J. (2009). Role of an artifact of dynamic algebra in the conceptualization of the algebraic equality. *Proceeding of CERME 6*, Lyon (Francia), www.inrp.fr/editions/cerme6

Chaachoua, H., Chiappini, G., Croset, M. C., Pedemonte, B., & Robotti, E. (2012). Introduction de nouvelles rerpésentations dans deux environnements pour l'apprentissage de l'algèbre. *Recherche en Didactique des mathématiques*, pp. 253–281.

Clements, D. H. (1999). Geometric and spatial thinking in young children. In J. V. Copley (Ed.), *Mathematics in the early years* (pp. 66–79). Reston, VA: NCTM.

DeThorne, L. S., & Schaefer, B. A. (2004). A guide to child nonverbal IQ measures. *American Journal of Speech-Language Pathology, 13,* 275–290.

Dehaene, S. (1997). *The number sense: How the mind creates mathematics*. New York: Oxford University Press.

Edyburn, D. (2005). Universal design for learning. *Special Education Technology Practice, 7*(5), 16–22.

Gallese, V., & Lakoff, G. (2005). The brain's concepts: The role of the sensory-motor system in conceptual knowledge. *Cognitive Neuropsychology, 22*(3–4), 455–479.

Geary, D. C. (1994). *Children's mathematical development*. Washington DC: American Psychological Association.

Geary, D. C. (2004). Mathematics and learning disabilities. *Journal of Learning Disabilities, 37,* 4–15.

Goldenberg, P., Cuoco A., & Mark, J. (1998). A role for geometry in general education, designing learning environments for developing understanding of geometry and space, pp. 3–44.

González, J. E. J., & Espínel, G. A. I. (1999). Is IQ-achievement discrepancy relevant in the definition of arithmetic learning disabilities? *Learning Disability Quarterly, 22*(4), 291–301.

Gracia-Bafalluy, M. G., & Noël, M. P. (2008). Does finger training increase young children's numerical performance? *Cortex, 44,* 368–375.

Griffin, S. A., Case, R., & Siegler, R. S. (1994). Rightstart: Providing the central conceptual prerequisites for first formal learning of arithmetic to students at risk for school failure. In K. McGilly (Ed.), *Classroom lessons: Integrating cognitive theory and classroom practice* (pp. 24–49). Cambridge, MA: MIT Press.

Heyd-Metzuyanim, E. (2013). The co-construction of learning difficulties in mathematics—teacher–student interactions and their role in the development of a disabled mathematical identity. *Educational Studies in Mathematics, 83*(3), 341–368.

Hittmair-Delazer, M., Sailer, U., & Benke, T. (1995). Impaired arithmetic facts but intact conceptual knowledge—Asingle case study of dyscalculia. *Cortex, 31,* 139–147.

Ianes, D. (2006). *La speciale normalità*. Erickson: Trento.

Ianes, D., & Demo, H. (2013). *What can be learned from the Italian experience?* (p. 61). La Nouvelle Revue de l'Adaptatione de la Scolarisation: Methods for improving inclusion.

Karagiannakis, G., Baccaglini-Frank, A., & Papadatos, Y. (2014). Mathematical learning difficulties subtypes classification. *Frontiers in Human Neuroscience, 8,* 57. doi:10.3389/fnhum.2014.00057

Karagiannakis, G., & Baccaglini-Frank, A. (2014). The DeDiMa battery: A tool for identifying students' mathematical learning profiles. *Health Psychology Review, 2*(4). doi:10.5114/hpr.2014.46329

Karagiannakis, G., Baccaglini-Frank, A., & Roussos, P. (2017). Detecting strengths and weaknesses in learning mathematics through a model classifying mathematical skills. *Australian Journal of Learning Difficulties*. doi:10.1080/19404158.2017.1289963

Kaufmann, L., Mazzocco, M. M., Dowker, A., von Aster, M., Gobel, S. M., Grabner, et al. (2013). Dyscalculia from a developmental and differential perspective. *Frontiers in Psychology, 4,* 516.

Kieran, C. (2006). Research on the learning and teaching of algebra. In G. Gutierrez & P. Boero (Eds.), *Handbook of research on the psychology of mathematics education. Past, Present and Future*. Rotterdam/Taipei: Sense Publishers.

Landy, D., & Goldstone, R. L. (2010). Proximity and precedence in arthmetic. *Quarterly Journal of Experimental Psychology (Colchester), 63,* 1953–1968.

Lagrange, J. B., Artigue, M., Laborde, C., & Trouche, T. (2003). Technology and mathematics education: A multidimensional study of the evolution of research and innovation. In A. J. Bishop & al. (Eds.), *Second International Handbook of Mathematics Education* (pp. 239–271). Dordrecht: Kluwer Academic Publishers.

Maddux, C. (1984). Using microcomputers with the learning disabled: Will the potential be realized? *Educational Computer, 4*(1), 31–32.

Mammarella, I. C., Giofrè, D., Ferrara, R., & Cornoldi, C. (2013). Intuitive geometry and visuospatial working memory in children showinsymptoms of non verbal learning disabilities. *Child Neuropsychology, 19,* 235–249. doi:10.1080/09297049.2011.640931

Mammarella, I. C., Lucangeli, D., & Cornoldi, C. (2010). Spatial working memory and arithmetic deficits in children with non verbal learning difficulties. *Journal of Learning Disabilities, 43,* 455–468. doi:10.1177/0022219409355482

Mariani, L. (1996). Investigating Learning Styles. *Perspectives, Journal of TESOL-Italy,* XXI, 2/XXII, 1, Spring.

Mazzocco, M. M. (2008). Defining and differentiating mathematical learning disabilities and difficulties. In D. B. Berch & M. M. Mazzocco (Eds.), *Why is math so hard for some children? The nature and origins of mathematical learning difficulties and disabilities* (pp. 29–47). Baltimore, MD: Brookes Publishing Company.

Mazzocco, M. M., & Myers, G. F. (2003). Complexities in identifying and defining mathematics learning disability in the primary school years. *Annals of Dyslexia, 53,* 218–253.

Mazzocco, M. M., & Räsänen, P. (2013). Contributions of longitudinal studies to evolving definitions and knowledge of developmental dyscalculia. *Trends in Neuroscience and Education, 2*(2), 65–73.

Michayluk, J. O., & Saklofske, D. H. (1988). Logo and special education. *Canadian Journal of Special Education, 4*(1), 43–48.

MIUR. (2011a). Dislessia, Gelmini presenta misure a favore di studenti con disturbi specifici di apprendimento (DSA) per scuola e università. Pubblicato online http://hubmiur.pubblica.istruzione.it/web/ministero/cs200711

MIUR. (2011b). Studenti con disturbi specifici dell'apprendimento. Rilevazioni integrative a.s. 2010–2011. Pubblicato online http://hubmiur.pubblica.istruzione.it/web/istruzione/prot5140_10

Mulligan, J. T., & Mitchelmore, M. C. (2013). Early awareness of mathematical pattern and structure. In L. English & J. Mulligan (Eds.), *Reconceptualizing early mathematics learning* (pp. 29–46). Dordrecht: Springer Science-Business Media.

Mussolin, C. (2009). *When [5] looks like [6]: A deficit of the number magnitude representation in developmental dyscalculia: behavioural and brain-imaging investigation.* Retrieved from http://dial.academielouvain.be/handle/boreal:21432

Nemirovsky, R. (2003). Three conjectures concerning the relationship between body activity and understanding mathematics. In N. A. Pateman, B. J. Dougherty, & J. T. Zilliox (Eds.), *Proc. 27th Conf. of the Int. Group for the Psychology of Mathematics Education 1* (pp. 103–135). Honolulu, Hawai'I: PME.

Nemirovsky, R., Rasmussen, C., Sweeney, G., & Wawro, M. (2012). When the classroom floor becomes the complex plane: Addition and multiplication as ways of bodily navigation. *Journal of the Learning Sciences, 21*(2), 287–323.

Noss, R., & Hoyles, C. (1996). *Windows on mathematical meanings: Learning Cultures and Computers.* In Mathematics Education Library: Kluwer Academic Publichers.

Núñez, R., & Lakoff, G. (2005). The cognitive foundations of mathematics: The role of conceptual metaphor. In J. I. D. Campbell (Ed.), *Handbook of mathematical cognition* (pp. 109–125). New York, NY: Psychology Press.

Papert, S. (1980). *Mindstorms: Children, computers, and powerful ideas.* NY: Basic Books.

Passolunghi, M. C., & Siegel, L. S. (2004). Working memory and access to numerical information in children with disability in mathematics. *Journal of Experimental Child Psychology, 88,* 348–367.

Piaget, J., & Inhelder, B. (1967). *The child's conception of space.* NY: W.W. Norton.

Piazza, M., Facoetti, A., Trussardi, A. N., Berteletti, I., Conte, S., Lucangeli, D. Dehaene, S., & Zorzi, M., et al. (2010). Developmental trajectory of number acuity reveals a severe impairment in developmental dyscalculia. *Cognition, 116*(1), 33–41.

Pinel, P., Piazza, M., Le Bihan, D., & Dehaene, S. (2004). Distributed and overlapping cerebral representation of number, size, and luminance during comparative judgments. *Neuron, 41*(6), 983–993.

Radford, L. (2003). Gestures, speech, and the sprouting of signs: A semiotic-cultural approach to students' types of generalization. *Mathematical Thinking and Learning, 5*(1), 37–70.

Radford, L. (2006). The anthropology of meaning. *Educational Studies in Mathematics, 61,* 39–65.
Raghubar, K. P., Barnes, M. A., & Hecht, S. A. (2010). Working memory and mathematics: A review of developmental, individual difference, and cognitive approaches. *Learning and Individual Differences, 20,* 110–122.
Ratcliff, C., & Anderson, S. E. (2011). Reviving the Turtle: Exploring the use of logo with students with mild disabilities. *Computers in the Schools, 28*(3), 241–255.
Resnick, L. B., Bill, V. L., Lesgold, S. B., & Leer, N. M. (1991). Thinking in arithmetic class. In B. Means, C. Chelemer, & M. S. Knapp (Eds.), *Teaching advanced skills to at-risk students* (pp. 27–53). SRI international.
Riconscente, M. M. (2013). Results from a controlled study of the iPad fractions game Motion Math. *Games and Culture, 8*(4), 186–214.
Robotti, E. (2014). Dynamic representations for algebraic objects available in AlNuSet: How develop meanings of the notions involved in the equation solution. In C. Margolinas (Ed.), *Task design in mathematics education. Proceedings of ICMI Study 22,* 1 (pp. 101–110). Oxford: ICMI.
Robotti, E. (2017). Designing innovative learning activities to face difficulties in algebra of dyscalculic students: Exploiting the functionalities of AlNuSet. In A. Baccaglini-Frank & A. Leung (Eds.), *Digital Technologies in Designing Mathematics Education Tasks—Potential and pitfalls*, (pp. 193–214). Springer.
Robotti, E., & Ferrando, E. (2013). Difficulties in algebra: New educational approach by AlNuSet. In E. Faggiano, & A. Montone (Eds.), *Proceedings of ICTMT11* (pp. 250–25). Italy: ICTMT.
Robotti, E., Antonini, S., & Baccaglini-Frank, A. (2015). Coming to see fractions on the numberline. In *Proceedings of the 9th Congress of European Research in Mathematics Education (CERME 9)*, Prague.
Rourke, B. P., & Conway, J. A. (1997). Disabilities of arithmetic and mathematical reasoning: Perspectives from neurology and neuropsychology. *Journal of Learning Disabilities, 30,* 34–46. doi:10.1177/002221949703000103
Russell, S. J. (1986). But what are they learning? The dilemma of using microcomputers in special education. *Learning Disability Quarterly, 9*(2), 100–104.
Santi, G., & Baccaglini-Frank, A. (2015). Possible forms of generalization we can expect from students experiencing mathematical learning difficulties. *PNA, Revista de Investigaciòn en Didàctica de la Matemàtica, 9*(3), 217–243.
Schmittau, J. (2011). The role of theoretical analysis in developing algebraic thinking: A Vygotskian perspective. In J. Cai & E. Knuth (Eds.), *Early algebraization a global dialogue from multiple perspectives* (pp. 71–86). Berlin: Springer.
Seron, X., Pesenti, M., Noël, M. P., Deloche, G., & Cornet, J. A. (1992). Images of numbers or when 98 is upper left and 6 sky blue. *Cognition, 44,* 159–196.
Sfard, A., & Linchevsky, L. (1992). Equations and inequalities: Processes without objects? *Proceedings PME XVI, Durham, 3,* 136.
Sinclair, N., & Pimm, D. (2014). Number's subtle touch: Expanding finger gnosis in the era of multi-touch technologies. *Proceedings of the PME 38 Conference*, Vancouver, BC.
Sinclair, N., & Zaskis, R. (in press). Everybody counts: Designing tasks for *TouchCounts*. In A. Leung, & A. Baccaglini-Frank (Eds.), *Digital technologies in designing mathematics education tasks potential and pitfalls*. Springer.
Stella, G., & Grandi, L. (2011). *Conoscere la dislessia e i DSA*. Milano: Giunti Editore.
Szucs, D., Devine, A., Soltesz, F., Nobes, A., & Gabriel, F. (2013). Developmental dyscalculia is related to visuospatial memory and inhibition impairment. *Cortex, 49,* 2674–2688.
Vamvakoussi, X., Dooren, W., & Verschaffel, L. (2013). Brief Report. Educated adults are still affected by intuitions about the effect of arithmetical operations: evidence from a reaction-time study. *Educational Studies in Mathematics, 82*(2), 323–330.
Vasu, E. S., & Tyler, D. K. (1997). A comparison of the critical thinking skills and spatial ability of fifth grade children using simulation software or Logo. *Journal of Computing in Childhood Education, 8*(4), 345–363.

Verschaffel, L., & De Corte, E. (1996). Number and arithmetic. In A. J. Bishop, K. Clements, C. Keitel, J. Kilpatrick, & C. Laborde (Eds.), *International handbook of mathematics education* (pp. 99–137). Dordrecht, The Netherlands: Kluwer.

Watson, S. M. R., & Gable, R. A. (2013). Unraveling the complex nature of mathematics learning disability: Implications for research and practice. *Learning Disability Quarterly, 36* (3), 178–187.

Wilson, A. J., Revkin, S. K., Cohen, D., Cohen, A. S., & Dehaene, S. (2006a). An Open Trial Assessment of "The Number Race", an adaptive computer game for remediation of dyscalculia. *Behav Brain Functions, 2*(20), 1–16. doi:10.1186/1744-9081-2-20

Wilson, A. J., Dehaene, S., Pinel, P., Revkin, S. K., Cohen, L., & Cohen, D. (2006b). Principles underlying the design of "The Number Race", an adaptive computer game for remediation of dyscalculia. *Behavioral and Brain Functions, 2*(1), 19. doi:10.1186/1744-9081-2-19

Zorzi, M., Priftis, K., & Umiltà, C. (2002). Brain damage: Neglect disrupts the mental number line. *Nature, 417*(6885), 138–139.

Part III
New Technological Spaces

Innovative Uses of Digital Technology in Undergraduate Mathematics

Mike O. J. Thomas, Ye Yoon Hong and Greg Oates

Abstract The ways in which digital technology is often used in university teaching of mathematics can be quite different from how it is employed in schools. This has the potential to form a discontinuity between school and university, making the transition less than smooth for students. In this chapter we consider several examples of how digital technology has been used with first year mathematics students in both New Zealand and South Korea. The approaches employed include: intensive use of technology, including formative and summative assessment practice; lecturer modelling and privileging of technology use; a versatile approach to calculus concepts that encourages epistemic exploration of local properties of functions; and novel orchestration of mathematical thinking through smartphone communication technology. We analyse each of these approaches using the theory of instrumental orchestration and outline some innovative aspects and benefits of them. The student perspective is also considered, with some evidence of the influence on student engagement and attitudes. We conclude by suggesting that in order to teach with digital technology in the manner described here good pedagogical technology knowledge (PTK) is required.

1 Introduction

In recent years there has been an increasing focus on research considering the mathematical transition from secondary school to university (see for example, Thomas et al., 2015). This research has highlighted a number of issues related to

M.O.J. Thomas (✉) · G. Oates
The University of Auckland, Auckland, New Zealand
e-mail: moj.thomas@auckland.ac.nz

G. Oates
e-mail: greg.oates@utas.edu.au

Y.Y. Hong
Ewha Womans University, Seoul, South Korea
e-mail: hongyy@ewha.ac.kr

E. Faggiano et al. (eds.), *Innovation and Technology Enhancing Mathematics Education*, Mathematics Education in the Digital Era 9,
DOI 10.1007/978-3-319-61488-5_6

transition. For example, one epistemic discontinuity is the change in emphasis from equality to inequality in the transition from algebra to analysis (Artigue, 2010). A second potential epistemic discontinuity, identified by Vandebrouck (2011), which is returned to in our discussion below, is the tendency for school mathematics to focus on pointwise and global perspectives of functions (see e.g., Fernández-Plaza, Rico, & Ruiz-Hidalgo, 2013), whereas university level mathematics requires a local perspective. While we have taken a slightly different standpoint below (see also McMullen, Oates, & Thomas, 2015) we agree with Vandebrouck's (2011) claim that often only point-wise and global perspectives on function are constructed at the secondary school. Students may evaluate functions, including derivatives, etc., at specific points and work globally on a function representation, such as applying a translation to a graph of a function, relating it to the algebraic formula $y - q = f(x - p)$ but may not consider the behaviour of a function on an arbitrarily small interval, such as $(x - \delta,\ x + \delta)$. This chapter considers a third area that could give rise to an epistemic discontinuity in the transition from school to university mathematics, namely the manner of use of digital technology (DT). In particular we consider how it might be used to address the provision of an environment in which students' might construct a local perspective on function.

2 A Potential Digital Technology Institutional Disjunction/Discontinuity

The use of DT in schools has been well researched, with a survey article by Lagrange et al. (2003) analysing 800 articles, leading to identification of seven dimensions related to ICT use in mathematics classrooms. The range of DT tools now available to teachers in developed countries is increasingly being recognised (Mousley, Lambdin, & Koc, 2003) along with the wide range of pedagogical opportunities some of these tools offer (Pierce, Stacey, & Wander, 2010). In general, studies have found small but significant, if not universal, positive effects from DT use in schools (e.g., Burrill et al., 2002; Cheung & Slavin, 2011; Graham & Thomas, 2000; Li & Ma, 2010) that do not hinder the development of mathematical skills (Ellington, 2003). However, Bressoud, Mesa, and Rasmussen (2015) report that although students were very comfortable using graphing calculators they were noticeably less comfortable doing calculations by hand. A key aspect in many school situations is that the students may have ready access to DT; especially through calculators, applets and more recently smartphone apps. A culture of investigative activity also exists in schools, with many studies describing ways to use technology such as graphics calculators and GeoGebra in an investigative manner (e.g., Goos, Galbraith, Renshaw, & Geiger, 2003; Pierce & Stacey, 2011).

On the other hand, the university situation can be more varied and problematic. The Reform Calculus movement promoted some major efforts in the use of technology (e.g., Bookman & Friedman, 1999; Meel, 1998; Park & Travers, 1996; Schwingendorf, 1999) and Thomas and Holton (2003) provide a comprehensive discussion of both theoretical and practical issues associated with technology use in undergraduate mathematics. Despite these efforts, use at scale remains sporadic, even within individual departments and is often centred on lecturers' research-oriented DT (Oates, 2011). Further, introducing new and complex technologies in a one-semester course raises issues of student instrumentation (Stewart, Thomas, & Hannah, 2005). A recent analysis of Calculus I programs (Bressoud, Mesa, & Rasmussen, 2015) showed mixed results regarding the use of technology at over 300 tertiary institutions, with lower use possibly due to some institutions valuing procedural fluency without technology. There are numerous studies describing individual approaches to the use of technology in undergraduate mathematics (e.g., Paterson, Thomas, & Taylor, 2011; Ng, 2011; Blyth & Labovic, 2009). However, such use is often either restricted to lecturer demonstration, or student use may be limited to use in tutorials and assignments in computer labs. Further, this use frequently exists in isolation from other courses in the department and may be more commonly associated with applied mathematics courses using specific technologies associated with research mathematics, such as Matlab, as opposed to other technologies which students maybe familiar with from school (Oates, 2011). For example, while Ng's (2011) study described the use of graphics calculators in an undergraduate calculus class, and Lin and Thomas (2011) have described the use of GeoGebra to develop student understandings of Riemann integration, such use is comparatively rare, much more common are studies such as those of Blyth and Labovic (2009), Tobin and Weiss (2011), and Paterson et al. (2011), all of which described the use of Matlab in applied courses such as differential equations and engineering mathematics.

Following on from the research observations above, several contributors to the Insights and recommendations from the national study of college calculus MAA report (Bressoud, Mesa, & Rasmussen, 2015) note aspects in the teaching of college calculus that support the innovative nature of the technology use described in this paper. For example, while use of graphics and CAS calculators are reasonably widespread in teaching and coursework, permission to use graphing calculators in examinations drops dramatically (Selinski & Milbourne, 2015), a sharp discontinuity between high school and college calculus. Larsen, Glover, and Melhuish (2015) describe two examples of innovative and ambitious teaching which mirror many of the elements incorporated in the studies reported here, emphasising that despite increasing use of technology, such examples of extensive and integrated approaches remain uncommon across institutions, and they support a more coordinated approach and more research into this field. At the post-calculus level, Rasmussen and Wawro (2016, in press) note that while both linear algebra, differential equations research studies are resulting in useful frameworks for interpreting student reasoning and planning for instruction, the role that technology can play in the modelling process is surprisingly missing and is an area ripe for future research. Hence, in line with these findings, this chapter considers a number of innovative approaches to using DT in the teaching of first year undergraduate mathematics courses that may suggest a better alignment with the school practice that students are accustomed to, while at the same time allowing them to see how DT can assist in building understanding of more advanced mathematical processes and concepts, such as a local perspective on function.

3 Instrumental Orchestration

The process of *instrumental genesis* involves the development of suitable mental schemes in order to transform a tool into an instrument suitable for the current task or activity. As writers such as Trouche (2004) have noted, this process is not automatic and requires teacher direction: "I introduce the term instrumental orchestration to point out the necessity (for a given institution—a teacher in her/his class, for example) of external steering of students' instrumental genesis" (Trouche, 2004, p. 296). Thus in instrumental orchestration a primary goal of a teachers' orchestrations is to help students engage in activity that will develop the appropriate mental schemes for producing techniques with either *epistemic value*, providing knowledge of the mathematical object under study, or 'productive potential', *pragmatic value* (Artigue, 2002). According to Trouche (2004) the orchestrations can act at several levels: that of the artifact itself; that of an instrument or a set of instruments; or a meta level, of the relationship of a subject with an instrument or a set of instruments. The nature of these orchestrations will also evolve over time in response to the development of technological tools. The orchestrations occur in the context of a *didactical configuration*, or an arrangement of the artefacts or tools in the learning environment, along with an *exploitation mode*, which is the "way the teacher decides to exploit a didactical configuration for the benefit of his or her didactical intentions" (Drijvers, Doorman, Boon, Reed, & Gravemeijer, 2010, p. 215). Orchestrations may be observed to comprise an "*intentional* and *systematic* management of artefacts, aiming at the implementation of a given mathematical situation in a given classroom" (Trouche & Drijvers, 2010, p. 676) or ad hoc decisions forming part of a *didactical performance* (Drijvers et al., 2010). This framework of instrumental orchestration has been suggested as a useful means of categorising observed teaching practices (Drijvers et al., 2010). More recent research (Drijvers, Tacoma, Besamusca, Doorman, & Boon, 2013), focused on an attempt to produce a taxonomy of whole-class and individual orchestrations, lists some of the latter as technical-demo, guide-and-explain, link-screen-paper, discuss-the-screen and technical-support (see the classification in Fig. 1).

Clearly the types of orchestrations in a didactical performance are dependent on the tasks used, and whether the techniques and utilisation schemes, employed in activity related to the task have an epistemic or pragmatic focus.

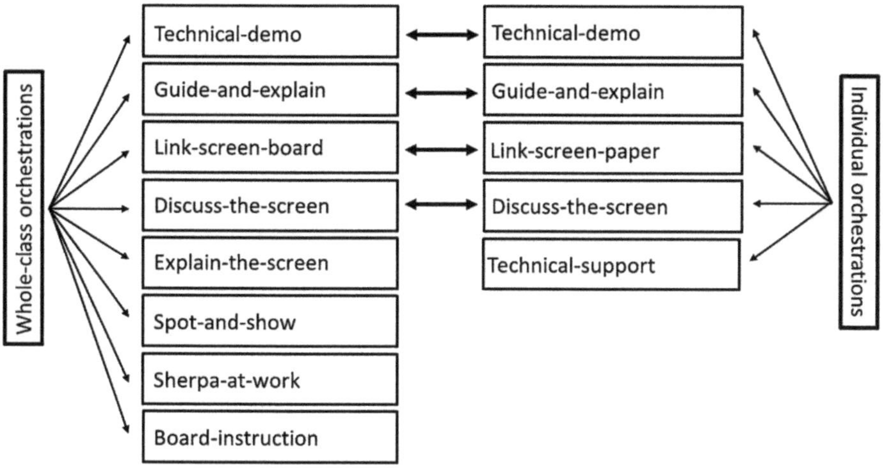

Fig. 1 Whole-class and individual orchestrations (from Drijvers et al., 2013)

4 Different Perspectives on Function

We now return to the ideas of Vandebrouck (2011), mentioned above, related to the various possible perspectives on function that are necessary in mathematical thinking. According to his analysis a pointwise property is one that only depends on the value of the function at a specific point x_0, such as evaluating the derivative of a function at a specific point, while a global property such as concavity is defined on an interval. In contrast, he considers a local property to be one that depends on the values of f in a neighbourhood of a specific point x_0, say $(x_0 - \delta,\ x_0 + \delta)$, as required for an understanding of continuity, something that is not often considered in secondary school. Thus if we consider two points $(x_0,\ f(x_0))$ and $(x_1,\ f(x_1))$ then, Vandebrouck's (2011) classification would mean that the average rate of change of the function on the interval $[x_0, x_1]$, $\frac{f(x_1)-f(x_0)}{x_1-x_0}$ is a global property, since it is defined on an interval. However, we have some issues with this classification based on contextual influence. For example, if we take the AROC of f over the interval $[x_0,\ x_0 + h]$, to be $\frac{f(x_0+h)-f(x_0)}{h}$ then is this still a global property irrespective of the size of h? Or does it become local as h becomes small, and if so at what point? Thus, in our discussion, we suggest a classification with four possible perspectives on function: global; interval; local; and pointwise. Of course, all of these involve a consideration of intervals of some size, ranging from one comprising a single point to the whole of the function's domain. We suggest that global thinking involves considering the function across the whole of its domain. Hence, the algebraic approach to global thinking will usually involve statements such as '$\forall x \in D$', where D is the domain of the function. Examples include even and odd functions (for an odd function $f(-a) = -f(a), \forall a \in D$), a continuous function (if we define it as continuous at every point in its domain), a periodic function (where, for some k, $f(a+k) = f(a), \forall a \in D$) and some transformations of functions (a translation, for example, where $g(a) = f(a-h)+k, \forall a \in D_f$). Similarly we will take a local perspective to involve intervals such as $[x-h,\ x+h]$ and $(x,\ x+h)$ where it is necessary to consider the behaviour of the function on intervals where a limit process is required, such as in this case, $h \to 0$. For a consideration of those intervals that lie between the macro and micro sizes of global and local we propose a designation of an interval perspective. Examples include the concavity of a function, intervals where the function is increasing or decreasing and intervals where $f(x) > 0$ (for some functions, of course concavity and increasing/decreasing are global properties—a quadtratic function exemplifies the former and a linear function the latter—but this is nor generally the case). The pointwise perspective considers the behaviour of a function at particular points in the domain, such as the sign of the gradient at $x = a$ or the value of the function at a point i.e., $f(a)$.

5 Intensive Use of Digital Technology

The first implementation of DT described here relates to the *Intensive Technology Innovation* study comprising one of three components of a wider research project, led by a research team at the University of Auckland, entitled *Capturing Learning in Undergraduate Mathematics*. The research involves a digital technology initiative in an entry-level mathematics course and follows a design experiment methodology where "a primary goal for a design experiment is to improve the initial design by testing and revising conjectures as informed by ongoing analysis of both the students' reasoning and the learning environment" (Cobb, Confrey, diSessa, Lehrer, & Schauble, 2003, p. 11). Theoretically, we apply an instrumental orchestration framework and seek to modify this through each stage of our interventions. In the course technology is employed in four major ways, as described in the course design principles below.

6 Course Design Principles

One guiding principle employed in the initial cycle of course design and construction, and one usually lacking in other research, and hence innovative, was that technology should be integral to the assessment process. Hence, each student was required to register and enrol into *MathXL*—a web-based homework, tutorial and assessment system, which was used for five skills quizzes that contributed 1% each to the final grade and the mid-semester test, which was worth 10%. The MathXL program allows for some measure of mathematical input and provides instant feedback by marking student answers. It also identifies topics where the student needs to focus their attention and directs them to sections in an online textbook as well as creating a personalised Study Plan. The quiz and test questions were largely free-response, exercising the MathXL facility for numerical, algebraic and graphical input of solutions, in contrast to static multiple-choice style questions. The quizzes comprised a time-limited, non-supervised assessment in which students had three attempts and their best score was recorded. The mid-semester test was also time-limited but held in a supervised computer lab with one attempt per question. Students were allowed access to CAS-calculators if they had them as well as access to all online resources, although time factors would have made this impractical for most. There is still some debate about whether the test should be technology-free (skills-based) or technology-active.

The second design principle in the study was that during the 36 one-hour lectures in the course the lecturers should model a range of appropriate technology since teacher-privileging of DT can be very influential (Kendall & Stacey, 2001). As Maschietto and Soury-Lavergne (Chap. The Duo “Pascaline and e-Pascaline”: An Example of Using Material and Digital Artefacts at Primary School) found, using several artefacts implies that teachers need to construct adequate orchestration of the different resources available. In our case a number of different DT platforms and programs were employed and so the lecturer was faced with similar orchestration issues. In the first implementation phase of the study DT included: a web-based graphing calculator; *YouTube* clips; applets to demonstrate critical features of mathematics; and mathematical websites. The second implementation phase continued most of this approach, with the exception of discontinuing the use of MathXL, but with greater use of the simple online graphing calculator *Desmos*, which was favoured by the students for its ease of platform access (on both computers and smart-phones) and its comparative instrumental and syntactic simplicity (Oates, Sheryn, & Thomas, 2014; McMullen, Oates, & Thomas, 2015). While MathXL does have many advantages in the power it gives students to pursue individual supported study, it is a commercial package with associated costs. Thus, in addition to the importance of teacher-privileging that influenced the overall approach, a main reason for this decision in the second phase was to minimise any disadvantage to students who did not have access to specific technologies. At the end of each lecture students were directed to webpages that illustrated the concepts at the heart of each topic, and a video-recording of each lecture was available to students within 24 hours via the learning management system. The third, somewhat innovative, design principle was that students were encouraged to use any technology platform they had access to, including all calculators, mobile phones, computers, tablets, etc. and any e-resources they could access with these, in lectures, tutorials, during home study and on other occasions, In other words, rather than seeing these DT resources as an obstacle to learning, preventing students from thinking deeply about the mathematics, as some would argue, they were viewed as having the potential to be turned into pragmatic and epistemic instruments, and hence of real value in learning.

The final design principle was that technology should be actively used in the one-hour weekly assessed tutorials that all students were expected to attend. In these a task with a context-based set of related questions was given to the students, who spent an hour working collaboratively, in groups of two or three, on the questions in the task. The students could use a computer with internet access and were also allowed to use any technological means to help them solve the tasks presented to them (i.e., Desmos, Matlab, Wolfram Alpha, etc.) including their own scientific or graphic calculators. Each group of students was required to hand in their answers to the questions at the end of the tutorial and their answers were marked as a group with the total mark produced from an equal weighting of group participation and mathematical correctness. The marks from the ten tutorials forms part of the assessment for the course and contributes a total of 13% towards the students' final grade.

Data was collected from many sources, and we will consider just five of these in the following discussion in order to evaluate the success of this approach. At the end of each iteration of the course students were asked to complete three questionnaires: a technology questionnaire; an attitude survey; and the standard university student course evaluation. Figure 2 shows examples of the questions used in the online technology questionnaire, which contained a mix of 19 open and closed questions, and investigated student use of technology in general; mathematics-focused technology use; and the student pattern of technology use during the course. The open questions had an unlimited response space. For the attitude survey, a Likert scale was constructed with five subscales in 29 randomised items, each with five possible responses (strongly agree, agree, neutral, disagree, and strongly disagree). The subscales measured: attitude to maths ability; confidence with technology; attitude to instrumental genesis of technology (learning how to use it); attitude to learning mathematics with technology; and attitude to versatile use of technology. The versatility subscale had four questions and the others five.

2. Do you think the lecturers made sufficient use of these technologies to help you understand their use and value? If not, specify which you would have liked more of.
3. Which technologies do you personally own or have easy access to? [list given]
5. Which mathematics learning technologies did you personally use in the course? Please indicate your frequency of use, and whether this was the first time you had used them.
7. What activities did you use technology for? Please specify which technologies you used for each of the following activities: [Lectures, assignments, tutorials, quizzes, other]
11. Describe the kind of activities you used technology for when working on mathematics problems in the course. [Open response]
14. Did you like the extensive use of technology in MATHS 102? Please explain.

Fig. 2 Examples of the open and closed questions from the questionnaire

Table 1 Examples of two attitude subscales

Learning mathematics with technology	Instrumental genesis
I like using technology to learn maths	Learning how to use technology is difficult for me
Using technology in maths is worth the extra effort	I work to improve my ability to use technology
Maths is more interesting when using technology	I often need to ask others how to use technology
Using technology hinders my ability to understand maths	I can understand a new technology as quickly as other people
I prefer working out maths by hand rather than using technology	Using technology wastes too much time in the learning of maths

In addition, there were five questions covering possible goals in technology use, which did not comprise a subscale.

Table 1 gives examples of some items from two subscales employed in data analysis in the study. In addition to the questionnaires, groups of volunteers were observed as they worked collaboratively in a computer lab on a series of rich technology-active tasks specially designed by one of the study team. Observation notes included the type of technology device used, when they were working or reading individually or all together, who in the group was not on task, and how the technology was used, for example modelling, checking answers, drawing graphs, performing algebraic manipulation or extending the task. Each of the students had their own CAS-calculators as well as the other technology available on the computer or their phones. One of the tasks focused on the average rate of change (AROC) of a function and its relationship to instantaneous rate of change, one of the key mathematical constructs targeted in the course. The link between these may be seen through the lens of our classification of perspectives on function, based on Vandebroucke (2011). For 'relatively large' intervals the AROC requires an interval perspective, but as the interval becomes very small and the limiting process is evoked, a local perspective is required to understand how it can lead to an instantaneous rate of change, which then has a pointwise focus. During lectures, the concept of AROC of a function was introduced using a *board-instruction* orchestration (see Fig. 1). In this mode, the lecturer wrote on paper, projected on to a large screen visible to all the students and recorded for the class lecture video.[1] The idea that a linear function and a polynomial through two points have the same AROC on the interval defined by those points was mentioned (see Fig. 3a). Basic AROC calculations were carried out using function notation and Fig. 3b shows an example where an interval perspective is used. Then the rate of change of a function at a point x_0, the derivative, was defined, using local thinking as the limit as $h \rightarrow 0$ of the AROC of the function on the interval $[x_0, x_0 + h]$. Rates of change were also emphasised for determining the nature of stationary points and for concavity (Fig. 3).

[1]All screenshots here are taken from the lecture videos provided to the students.

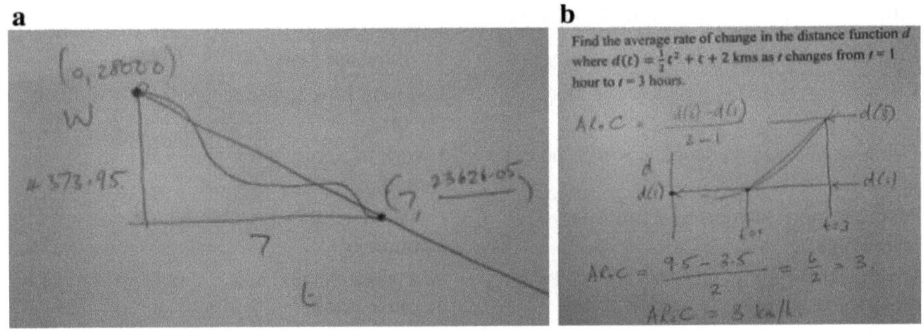

Fig. 3 Two screenshots taken from lecture videos

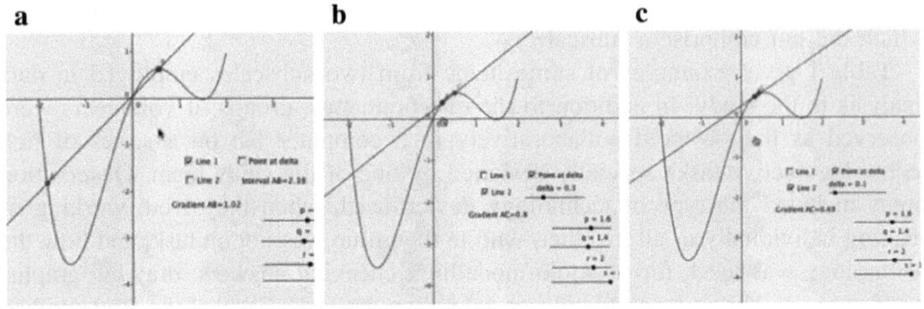

Fig. 4 Screenshots showing dynamic use of GeoGebra for interval and local AROC

7 Whole Class Instrumental Orchestration

During the introduction of AROC a program, written by the lecturer using GeoGebra, was displayed (see Fig. 4). Using dynamic dragging with sliders, and an *explain-the-screen* orchestration, the lecturer was able to promote both interval and local thinking when presenting examples of the AROC between two points, either a variable or a fixed distance apart, and link the screen view to mathematical constructs.

Some of these examples (Fig. 4a) could be said to illustrate interval properties due to the macro size of the distance between the points, while others were local properties, with a small delta, down to 0.1 (Fig. 4b and c). It did not prove possible to provide this program for students to engage with during the lecture, which would have been valuable, and in line with the kind of constructionist principles described by Kynigos (Chap. Innovations Through Institutionalized Infrastructures: The Case of Dimitris, His Students and Constructionist Mathematics). However, it was made available to them through the learning management system with the encouragement to experiment outside of class. Two further examples of *technical-demo* orchestrations using the web-based Desmos graphing program are shown in Fig. 5. Here

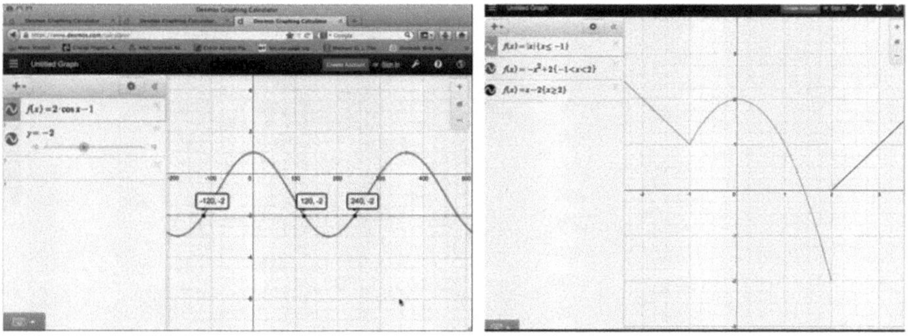

Fig. 5 Screenshots showing use of explain-the-screen with Desmos

the left hand screen shows a demonstration of a technique for finding and displaying approximate solutions to equations for $2\cos(x) - 1 = -2$. The right hand screen shows a possible use of Desmos to draw functions with split domains.

Students were constantly encouraged to change and extend the examples given in the lectures by investigating for themselves what the program response to various inputs would be, and 50% of the questionnaire respondents said that they used Desmos during the lectures. We see this kind of orchestration that usually followed a *technical-demo* as a new development of the Drijvers et al. (2013) classification, which we have called a *guide-to-investigate*, with students immediately encouraged to use Desmos, or other technology in their possession, to investigate further examples.

The final program that featured through the course was Wolfram Alpha. In Fig. 6 we provide three screen shots showing two examples of its use.

All three screens were employed in *explain-the-screen* orchestrations. The top screen shows how we can find the local maximum value of a function. Although this is a black box process, one advantage of Wolfram Alpha here is that it makes links between the algebraic and graphical representations (Thomas, 2008). The other two screens enabled discussion of a valuable DT technique that employs the absolute value of a function for finding the area between a function and the x-axis, and why a difference may occur between this area and the 'standard' definite integral.

An active-technology tutorial task, in a financial context, was designed around these lectures on the AROC concept. Students in the tutorial were asked to engage with and respond to a description of two mock students answering a problem associated with the graph of $f(t)$, as shown in the condensed excerpt of this task (without the associated diagrams) in Fig. 7.

For our final source of data, we examined student responses to a question included in the final examination, which examined their understandings of AROC. Unfortunately, one of the limitations of this study was that university policy on internal examinations limited student access to technology to CAS or graphics

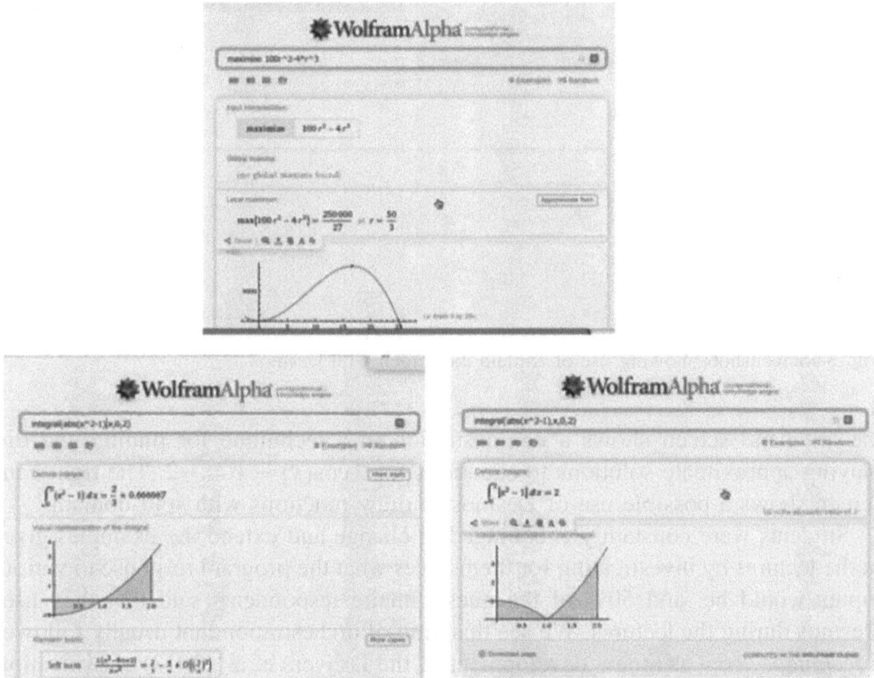

Fig. 6 Screenshots showing use of explain-the-screen with Wolfram Alpha

> …Raj said this was an interesting problem but a bit difficult. He found the graph of the function $f(t) = 0.025(2\sin(t) + t\sin(2t) - 2t\sin(3t) + 65)$, $0 \leq t \leq 25$, [graph given] which looked a bit like one of these graphs. He suggested that they work together to find out for this graph which *t* interval of size 2 has the greatest average rate of increase. It didn't take Sonja long to suggest a method. She said "You take the point at which the rate of change is greatest and take a *t* interval of 1 either side of it." What do you think of Sonja's method? Is she right? Investigate the greatest average rate of increase over a *t* interval of 2 for this graph. Where does it occur? If the *t* interval is 1 instead, where does greatest average rate of increase occur then? If the *t* interval is k instead, where $k \geq 0.5$, for what value of k does the greatest possible average rate of increase occur? If the *t* interval is k again, what happens to the average rate of increase as k gets smaller and smaller, i.e. as $k \to 0$? Describe in detail a method that would help Raj and Sonja find greatest average rates of change for graphs like this one.

Fig. 7 Technology-active tutorial task based on the AROC

calculators. We recognise that this may affect any conclusions that can be drawn from this final source of data.

It is important to recognise that the kind of instrumental orchestrations that can be employed are very much dependent on the pedagogical technology knowledge

(PTK) of the lecturer (Thomas & Hong, 2005; Hong & Thomas, 2006; Thomas & Palmer, 2013). In this case the lecturer in the second implementation was very experienced in using technology in teaching, having over 25 years experience, a high level of instrumental genesis with a number of DT tools, positive attitudes to technology use, a good mathematical background and strong confidence in using technology in mathematics teaching.

8 Student Technology Use

With respect to the questionnaires, student responses to technology use in each phase were consistent (Oates et al., 2014), although there was some evidence in the second implementation that the sustained intensive technology approach might be leading to greater usage, for example 100% of respondents in the second phase reported using Desmos in the course compared with 85% in the first. It also seems teacher-privileging may have had led to an increased use of GeoGebra with 10% in phase 2 compared to only one student in phase 1. Desmos was easily the most popular platform (80% very useful or useful; only 10% had never used it), with usage in lectures (50%), assignments (77.8%), tutorials (77.8%), and quizzes (77.8%). The students particularly liked its ease of access and use: "very easy to use and very easy to access", "useful as it is very responsive (quick) and extremely easy to use" and "Easy to use". In terms of lecturer modelling, 91.7% said the lecturer made sufficient use of technology in the course (cf. 76.9% in 2014), with 90.9% affirming that they received sufficient help with the technology ("Lecturer always explains"). Further, 83.3% liked the extensive use of technology in the course ("The use of technology was great, seeing the graphs and how they work in Desmos was really useful"; "It provides another perspective when solving problems"; "Yes, it's nice to know that we are moving with the advancement in technology") and 91.7% thought the technology helped in their learning of mathematics, for example, helping to visualise solutions ("Graph is much easier to understand and solve problems").

The attitude survey demonstrated, with reasonable reliability (supported by Cronbach Alpha measures), that students at the end of the course had positive attitudes towards technology to learn mathematics, to learn the techniques and construct the schemes required to do so, and a confidence to follow through on both. The subscale for *Attitude to Learning Mathematics with Technology* had a mean response of 3.24/5 (Cronbach alpha 0.71), *Confidence with Technology* a mean of 3.69/5 (Cronbach alpha 0.77), *Attitude to Instrumental Genesis* a mean of 3.62/5 (Cronbach alpha 0.64) and *Attitude to Mathematics Ability* a mean of 3.58/5 (Cronbach alpha 0.86).

During the DT-active tutorial task, technology was very much embraced by the students, with all groups, after reading through the task, immediately using Desmos to plot the function given to them: $f(t) = 0.025(2\sin(t) + t\sin(2t) - 2t\sin(3t) + 65)$, $0 \leq t \leq 25$. They then zoomed in on aspects of the graph they were interested in or plotted it against other functions they came up with. Scientific calculators were also used but most of their work was done with Desmos. Students also used their own course or internet resources, such as Wolfram Alpha, to look up mathematical concepts that they did not remember or could not find in their notes (possible reasons for this are discussed later), such as AROC. While the technology was well used, the students in the closely observed focus group tended to perform by-hand calculations, integrated with computer use, retrieving data or ideas and moving back and forth between the two environments. There was considerable discussion between the two of them. It was apparent that they knew how to calculate AROC:

> A: So you work out the average rate of change between that point and that point which is going to be 3.2 take away 0.1, which is pretty much that bottom point there. Between those two. And there's only a difference of one. So you've got an average rate of change of 3.1. Are we good on that?

They also demonstrated some idea of interval properties, and the effect on AROC of reducing the interval size, in essence exhibiting the ability to shift from interval thinking to local thinking involving limiting values.

> A: So that will give you the steepest line there. The other one is that one, which is pretty close, between the 29th and 12 o'clock on the 29th. But it's not quite as good. But as your k gets smaller, so as your k interval gets smaller and smaller and smaller, that one will become your steepest line. But then it will swap to that one.

> A: …so m gets smaller and smaller…As m gets smaller, the greatest rate of change is going to effectively be steeper. Until you get to the stationary points. So the stationary points will remain the same, but as you get closer and closer…

The lectures and tutorial task were followed up with the following question on AROC set in the final examination:

> The London Eye (picture provided in exam) is a giant circular ferris wheel in London, UK. The height, H metres, of passenger capsule A above the centre of the wheel t hours after the wheel starts to move is given by: $H(t) = 60\sin\left(4\pi t + \frac{\pi}{4}\right)$. What is the average rate at which capsule A is rising during the period from $t = 0$ to $t = \frac{1}{16}$ hours?

While technology was not essential to answer this question, the removal of the customary technology such as Desmos, which observations had already suggested many students turn to almost automatically, may have influenced the results. Regardless of the reasons, the question proved relatively difficult, with only 21.6% of the students fully correct and 62.5% gaining no marks out of 2 on part (iv) of the question. The students were comfortable using the function notation $H(0)$ and $H\left(\frac{1}{16}\right)$, although few gave the exact answer (even when close to it), resorting, not surprisingly, to calculators to work out the answer (Fig. 8).

Overall this task, which was written with active technology use in view, generated a lot of discussion among the students and they investigated this task in more depth than they did previous tutorial tasks that had not been designed specifically around the use of technology. However, the progress of some students was limited by their desire to employ Desmos, due to its relative ease of use, rather than other programs such as GeoGebra that would have allowed a greater array of techniques

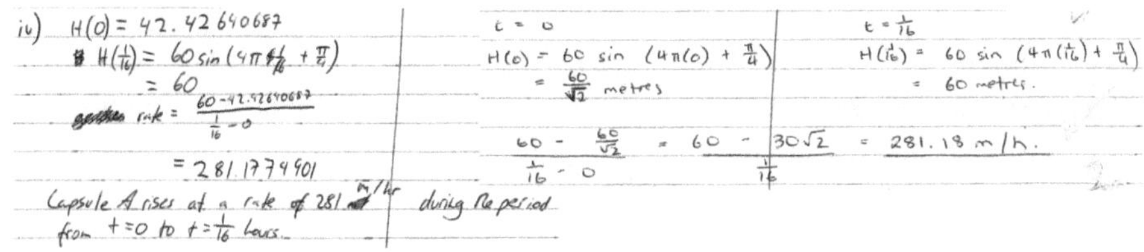

Fig. 8 Sample working on the examination question on AROC

to be employed in activity on the task. In this respect, it seemed evident that the greater use of Desmos in the second phase of the study by the lecturer had influenced students' propensity to use Desmos, rather than explore other technologies, as was more the case in the first phase of the study (McMullen, Oates & Thomas, 2015).

9 New Orchestrations with Smartphone Digital Technology

The constant evolution of digital technology presents teachers with the question of how to innovate in order to adopt new technologies with learning potential into teaching in an interactive manner that will foster mathematical thinking (Thomas, Monaghan, & Pierce, 2004), and exactly how it might be orchestrated. This requires consideration of questions such as, what interactions are desirable between student, teacher and technology in order to promote mathematical thinking and understanding, and how can the teacher orchestrate these?

One crucial aspect of the rapid development of DT tools is the role of connectivity in establishing collaboration through communication (Hoyles et al., 2010). This has often been interpreted as existing within the didactical configuration, although the implications of cloud technology have also been described as a 'revolutionary manifestation' of connectivity (Trouche & Drijvers, 2010).

The innovative potential of the latest mobile technology devices, such as smartphones, for learning has been recognised (Coffland & Xie, 2015), based on their accessibility, immediacy and portability. Although research is still in its infancy, possible uses, such as dynamic demonstrations of concepts, access to expert advice on problem solving methods, immediate feedback on practice problems, and dynamic concept exploration have all been suggested (Coffland & Xie, 2015). Other possibilities for as yet untapped usage, presented by White, Booker, Carter Ching, and Martin (2011), include the opportunity for students to make mathematics personal by allowing them to capture their real problem-solving situations by means of pictures, audio, and videos. While there have been some research studies using mobile technology, including: practicing skills related to a given content area (O'Malley et al., 2013); collaborative learning and spontaneous reaction (Norris, Soloway, Tan, & Looi, 2013); and investigation of how socio-cultural and situated learning aspects are reflected in learning experiences (Genossar, Botzer, & Yerushalmy, 2008), further studies related to learning are needed. In particular, there have been few, if any, studies that have considered the use of mobile technology to learn mathematics in the university setting, making this an innovative area.

Young people today form a strong bond with their mobile phones and there seems little doubt that the assemblage of students and their mobile phones (de Freitas, Ferrara, & Ferrari, Chap. The Coordinated Movements of a Learning Assemblage: Secondary School Students Exploring Wii Graphing Technology), is a complex area worthy of study. However, while the availability of smartphones is opening up even greater opportunities for innovative implementation and communication in mathematics learning at the tertiary level, this brings with it challenges for teacher orchestration. In a recent study we examined smartphone use with 134 students in two pre-calculus classes of the same university course in Korea. This considered the dynamic, immediate and communicative aspects of the technology in students' learning and problem solving. The course was taught using lecturer demonstration with GeoGebra, Geometer's Sketchpad and graphic calculator apps on a smartphone, which the students downloaded during the class. Students also used *KakaoTalk* on the SNS (Social Network Service), which allows one to send and receive messages and pictures on the screen of a smartphone, making immediate feedback during a class possible. The researcher demonstrated how to download KakaoTalk and put a copy on their e-class (electronic learning) website which was accessible to enrolled students. Once again, the lecturer here had excellent PTK (Thomas & Hong, 2005; Hong & Thomas, 2006). She is very positive in her attitude to technology use in mathematics teaching, has over 20 years experience with technology in teaching, along with high level of instrumental genesis using a number of DT tools, is very confident and has a good mathematical background.

Examples of the kind of activity this lecturer uses to encourage global, interval, local and pointwise thinking about function are included here to provide a context for the examples below (see Hong & Thomas, 2015 for further details). Having defined a rate of change function $r(h) = \frac{f(2+h)-f(2)}{h}$ for $f(x) = x^2$ and generated numeric approximations for r, her aim is to move to a general rate of change. A CAS calculator can be used to obtain the derivative at $x = a$ by defining a function $slope(h) = avgRC(f(a), a, h)$, $a = \{-1, 0, 1, 2, 3\}$ as the average rate of change over an interval of size h. As in the study above this can initially focus on interval thinking but lead to local thinking as the intervals get progressively smaller. Thus students can investigate the change of slope and, by taking the limit, see that it gets close to $\{-2, 0, 2, 4, 6\}$, and hence conjecture that the derivative is $2x$ (see Fig. 9).

Next we illustrate the kinds of novel orchestrations that she was able to introduce. In this course the students were employing their smartphones as a graphic calculator as well as a communication device. In Fig. 10 we see a student communicating with the teacher through KakaoTalk on the smartphone. The second

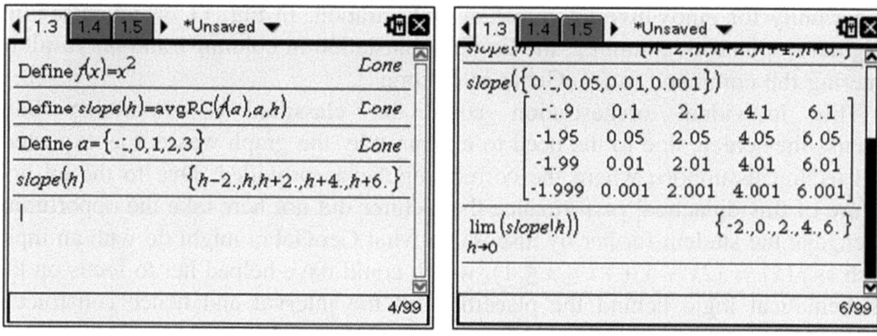

Fig. 9 Calculator screens showing interval slope calculation

Translation

Student: Sir, I am going to sketch the conditional graph using GeoGebra, it is cut out when I put $2x - 1(-1 \leq x \leq 1)$. How do I define the interval, please?

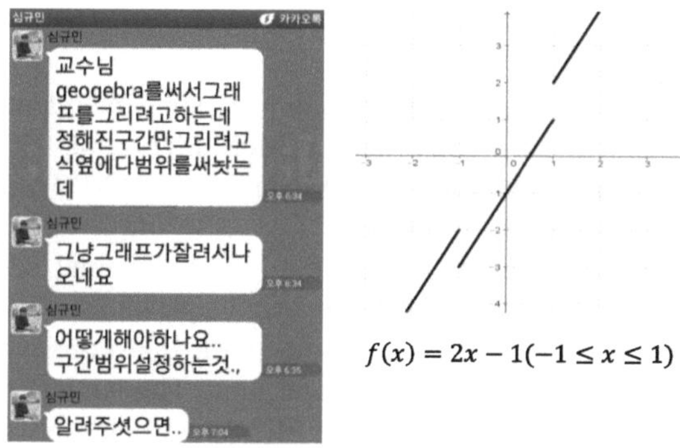

Fig. 10 A student utilising KakaoTalk to communicate with the lecturer

column contains the original communication in Korean, the first an English translation and the third the student input in GeoGebra and the resulting graph.

Entering $f(x) = 2x - 1(-1 \leq x \leq 1)$ into GeoGebra the student was surprised by the discontinuous graph obtained, what she called 'cut out'. Realising this was incorrect since she wanted the graph of $2x - 1$ on the interval $[-1, 1]$ to display, she had asked how to define the interval correctly. It appears that solving her dilemma involved a transition from interval thinking about function (why did she have a different definition on each interval) to a global perspective (the intention was that the function be defined on the interval $[-1, 1]$). Here the student knew what the function should look (i.e., the global view) like but lacked the instrumentation needed to produce it. We see that GeoGebra has behaved in a subtle manner, taking the input for f to mean subtract 1 from the function $2x$ only on the interval $[-1, 1]$ and leave $2x$ elsewhere. Interestingly, and illustrating the potential vagaries of DT techniques, and the associated difficulties for instrumental genesis, the format of this command with {} brackets would produce the desired effect in the Desmos program. However, the communication technology provided the teacher with the opportunity for innovative instrumental orchestration. In Fig. 11 we again see her original response in column 2, the English translation in column 1 and the result of entering the command in GeoGebra in column 3.

This individual orchestration could be classified as involving both discuss-the-screen, due to the need to explain why the graph was not as expected, and technical-support, where the correct input was provided. Due to the ad hoc nature of this didactical performance the lecturer did not here take the opportunity to engage the student further by discussing what GeoGebra might do with an input such as $f(x) = (2x - 1)(-1 \leq x \leq 1)$, which could have helped her to focus on the mathematical logic behind the placement of the interval and hence construct a suitable scheme for using them. In this latter case GeoGebra keeps the domain as $\mathbb{R}$ and draws the graph of the function on three intervals

$$f(x) = \begin{cases} 2x - 1 & -1 \leq x \leq 1 \\ 0 & \text{otherwise} \end{cases}$$

rather than restricting the domain to $[-1, 1]$ as intended. This type of orchestration, which, as mentioned above and in McMullen et al. (2015), does not appear to be covered by the taxonomy of Drijvers et al. (2013), and could, we believe, be classified as *guide-to-investigate.*

In a second example, seen in Fig. 12 (original screenshots with an English translation underneath), the student sends questions while using a GC to work on an assignment and the lecturer responds to the questions with a didactical performance executed through KakaoTalk in real time. The student asks whether her method is correct or not, and to make sure of her working she sends the screenshot of the graph (see the enlargement in Fig. 11). Invoking a pointwise view of function to find the points where the function has the value zero, the student tries to solve the equation $k^2 - 6k + 13 = 0$ algebraically but is unable to and wonders if she has made an error. Once told that there are no solutions she realises that she may not have 'to use the quadratic formula for the roots' and wonders how she might verify

Translation

Lecturer: Did you solve your problem of the interval?

You have to enter the following in the input window: $if(-1 \leq x \leq 1, 2x - 1)$

Student: That's what I wanted to know.

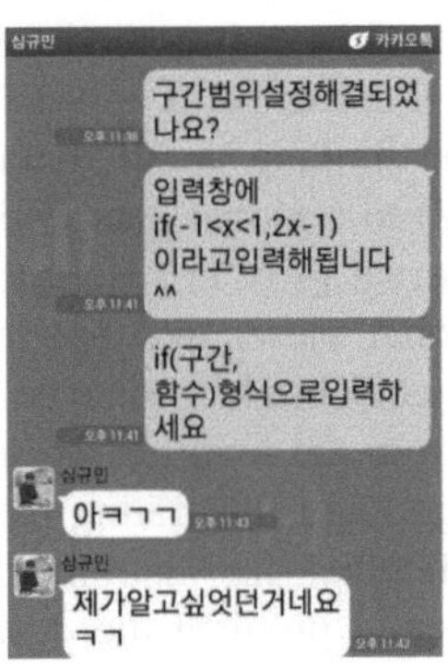

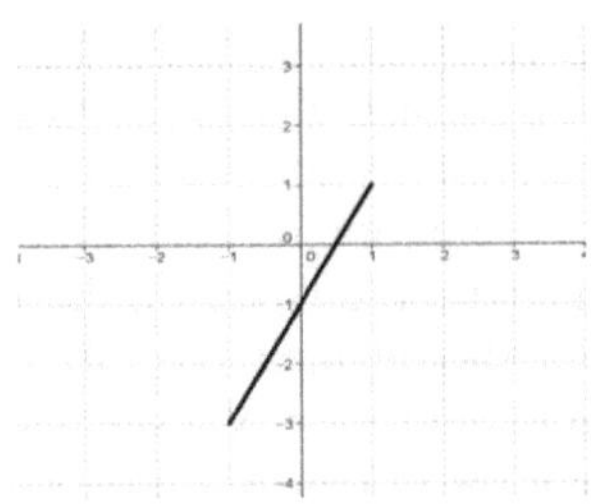

$f(x) = if(-1 \leq x \leq 1, 2x - 1)$

Fig. 11 The lecturer response using KakaoTalk for instrumental orchestration

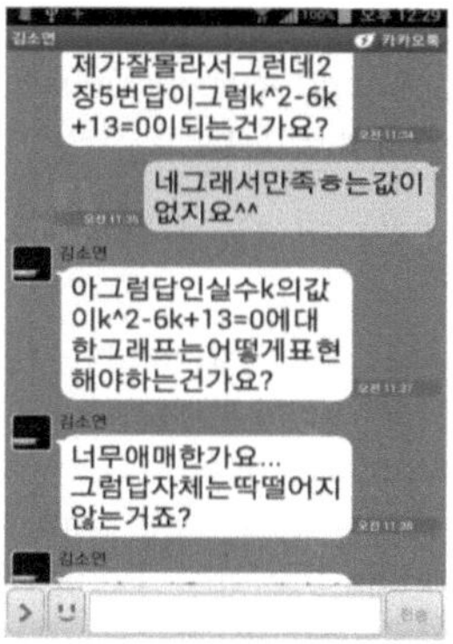

Student: The answer to question 5 in chapter 2, is *k*^2–6*k*+13=0, isn't it?
Lecturer: Yes, so a value satisfying this does not exist.
Student: How do I represent the graph of *k*^2–6*k*+13=0? The answer doesn't look clear. The value of *k* doesn't have an exact value, right?

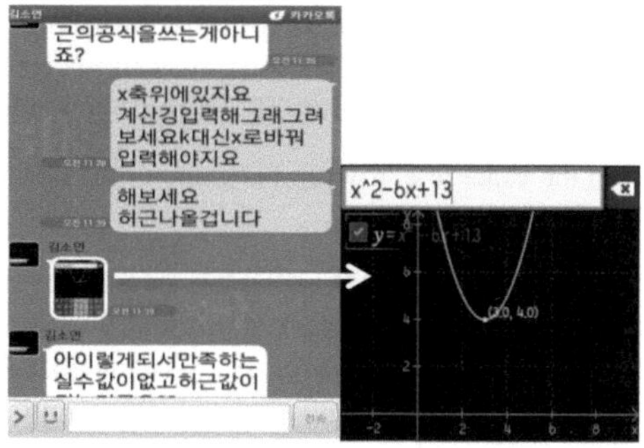

(enlarged graphic calculator screenshot on the right)
Student: Then, I don't have to use the quadratic formula for the roots?
Lecturer: To see the status of *k*, sketch the graph of *k*^2–6*k*+13 for *k*, you have to change it to *x*^2–6*x*+13 instead of *k*. Try it. Then you can see that the value of *k* does not exist on the *x*-axis.
Student: I see, I understand why I don't have real roots looking at the graph.

Fig. 12 The student and lecturer discussion using KakaoTalk

this using what she describes as 'the graph of $k^2 - 6k + 13 = 0$'. The lecturer, who realises that the GC technique requires a function in x, orchestrates the student's instrumentation by suggesting drawing the graph of the function f, where $f(x) = x^2 - 6x + 13$. The student follows this advice, and then sends the resulting graph through the communication channel with the comment "I see, I understand why I don't have real roots looking at the graph". The change of representation has enabled the GC to provide epistemic insight, including a change in focus from pointwise to global; seeing that the function's graph lies above the x-axis across the whole of its domain.

In this case the lecturer's orchestration could be described, firstly, as technical-support, assisting the student to see that the GC will only plot graphs in terms of x not k. This has pragmatic value, since it leads to the graph production. However, at a deeper level the orchestration is helping the student develop an appropriate mental scheme with genuine epistemic value. It has the potential to produce the knowledge that the particular variable used in a function is irrelevant, leading to a technique whereby it may be substituted by any other variable. This has many crucial applications in mathematics, such as in the Fundamental Theorem of the Calculus. The second aspect of the orchestration here is the encouragement to experiment in order to learn ("Try it. Then you can see that the value of k does not exist on the x-axis"). In this case it involves having the versatility to link the function across two representations, with the mathematical outcome much easier to see from the graph than the algebra, and, again, could be classified as *guide-to-investigate*.

The third example of communication and orchestration afforded by KakaoTalk presented here exemplifies a discuss-the-screen orchestration, and it occurred while the students are using the GC in the class while working on the task:

> Find the value of k, if when the graph of the function $y = \frac{x-1}{x-2}$ is reflected onto itself in the line $y = -x + k$.

This task combines several perspectives on function. The students were encouraged to solve the task graphically using the GC and then send a screenshot of their working to the lecturer via KakaoTalk. Thus the lecturer was able to see what students are doing, check their answers and provide necessary feedback immediately, using *guide-and–explain* or *explain-the-screen* orchestrations. Figure 13 shows an example from one particular student who has sent the graph of $y = \frac{x-1}{x-2}$, as well as those of the linear functions $y = -x + 3$, $y = -x + 2$ and $y = -x + 5$. In this case, to assist the student to find the solution, k = 3, the lecturer took the following steps. She asked her to compare the graph of $y = \frac{x-1}{x-2}$ with that of $= \frac{1}{x}$, and, after seeing the graphs, the student, employing a global view of the function, answered that it is translated about 2 units along the positive x-axis and shifted up 1 unit. The lecturer then asked where the horizontal and vertical asymptotes intersect and the student answered that they intersect at the point (2, 1).

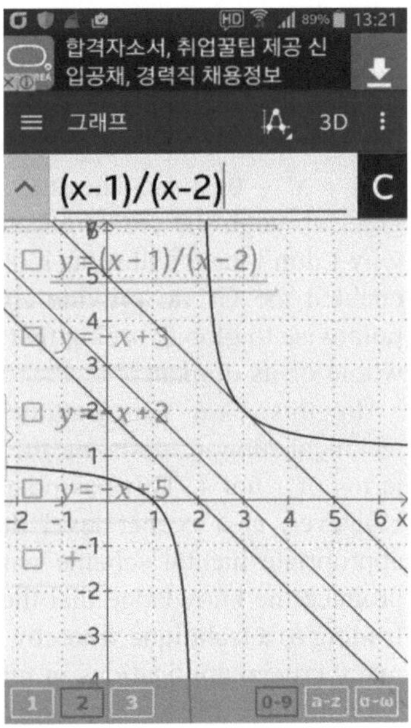

Fig. 13 Communication between a student and the lecturer using screenshots

Following this the lecturer asked through which point the line of reflection of the graph of $y = \frac{x-1}{x-2}$ must pass, and the student, using the pointwise idea of an invariant point, answered (2, 1). After this exchange the student was able to answer that the appropriate linear function is $y = -x + 3$. Some confirmation of the epistemic value of the DT approach was obtained from the mid-term test, where this example was extended to finding a range of possible values for k, as follows: The hyperbola $y = \frac{x-1}{x-2}$ intersects the line $y = -x + k$ at two points. Find the range of possible values for k and sketch the graphs. This question involves a global perspective, with its vertical translation of the straight line. The results support the hypothesis that a DT 'guide-to-investigate' orchestration may be effective in developing the students' visual thinking and conceptual understanding. In this case, 60.4% of the students used an algebraic method with the discriminant to find two intersection points, followed by a graphical method where they employed a translation for $y = -x + k$ (see one example in Fig. 14).

Our final example of smartphone communication and orchestration occurred when students were working on the following task.

> It is known that a population of a certain of moth changes on an annual cycle, which can be approximately modeled by the function: $P(m) = m^2 - 8m + 18$, where m is the number of the month ($0 \leq m \leq 11$) and P(m) is in thousands. In what month is the population at a minimum, and what is this population?

As seen in Fig. 15, the students used the GC to sketch the graph of $P(m) = m^2 - 8m + 18$, and then transferred their graphical work to paper. Here the lecturer has communicated the alternative version of the function, completing the square to give a version of $P(x) = (x - 4)^2 + 2$.

The students have quickly picked up the epistemic value of this orchestration as a route to the solution, without the need to differentiate the function. We see that both students A and B have completed the square and recognised that the minimum value is 2, occurring at $m = 4$, writing $P(4) = 2$.

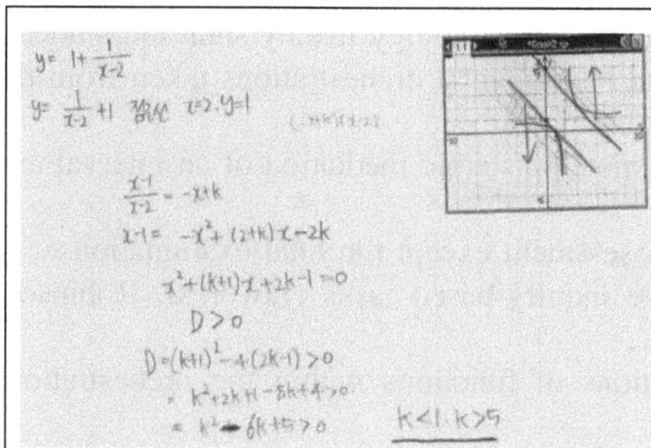

This student algebraically simplified to find the asymptotes $x = 2$ and $y = 1$, and found the range for k using D>0 from $\frac{x-1}{x-2} = -x + k$. The value of k was found from the intercept points on the y-axis from $y = -x + k$ and the range of k considered that $y = -x$ was translated by k units. Also, the graphical translation was shown on the graph by the direction of arrows.

Fig. 14 A student's solution integrating algebraic and graphical techniques

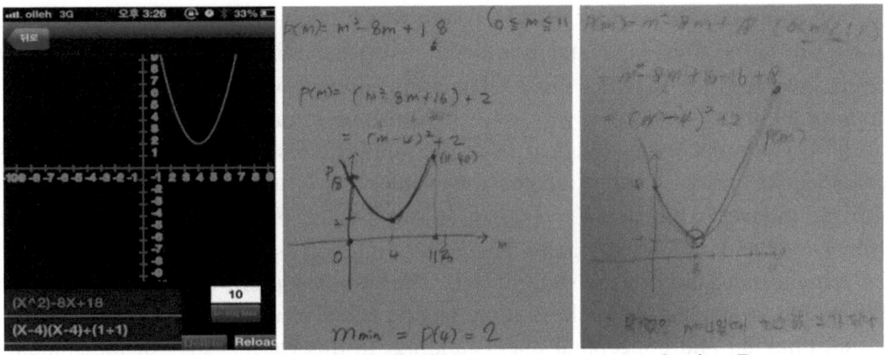

GC Student A Student B

Fig. 15 Students' working with the graphing calculator and KakaoTalk

Hence, although many students would naormally solve this problem in a standard by-hand manner using differentiation techniques, the GC, along with lecturer orchestration we can describe as *guide-to-maths-technique*, led these students to link the graphical and algebraic representations in a versatile manner (Thomas, 2008) through the technique of completing the square. In addition, the standard method focuses on pointwise thinking to find the points where the derivative is zero, but completing the square and drawing the graph shifts the focus to a global one where the student can see that across its whole domain the function is greater than the value at the vertex.

10 Conclusion

In this chapter we have described some recent research involving innovative aspects of technology use in a university mathematics environment that include:

- Integrated technology involving intensive technology use by staff and students
- Courses deliberately framed around instrumental orchestrations taken from the recent literature (Drijvers et al., 2013)
- Course design that used DT to promote epistemic mediation of an interval and local perspective on function
- DT integrated into all aspects of assessment except the final examination
- Small group activity on DT-active inquiry-based tasks (Jaworski, Robinson, Matthews, & Croft, 2012)
- Dynamic DT multiple representations of functions within the orchestrations framework
- Dynamic, immediate use of mobile technology in a university problem solving context

- Use of mobile technology for instant, direct communication between students and lecturer involving mathematical output and thinking
- Linking of interval and local perspectives of function.

In both of the studies reported on here, and in other examples (Hong & Thomas, 2015), we have seen that it is possible to design and construct a teaching approach at university level using DT that can assist students to construct both interval and local perspectives on functions. In this way one existing discontinuity between school and university-based epistemologies may possibly be satisfactorily addressed.

At a practical level, this included integrating technology into as many aspects of the course as possible, thus mirroring and even extending the use of technology that most students now bring with them from school, such as encouraging the use of Desmos and Wolfram Alpha in the first study, and KakaoTalk in the second. Staff modelled the use of technology in lectures, encouraged students to engage actively with it in lectures and provided examples of multiple different platforms for students to explore outside of class. Other aspects included overt use of technology in assessment (e.g., course tests and compulsory use in assignments), and specially-designed active technology tasks for use in small-group tutorials. Observations and survey responses suggest that students both enjoyed and benefited from this approach, with evidence that the tasks promoted students' exploration using the multiple representations provided by the technology. The first study supports the value of both the concept of instrumental orchestration proposed by Trouche (2004) and teacher privileging (Kendal & Stacey, 2001), with a clear indication that the teachers' promotion of technology (especially Desmos) and the examples used in lectures and developed in the tasks guided students' choices and uses of technology.

The first study also provided an indication that deliberately designing a course around the instrumental orchestrations framework (Drijvers et al., 2013) may lead to opportunities for additional orchestrations beyond those previously described. Tutorial observations suggested that the use of Desmos with the AROC problem in lectures, incorporating specific orchestrations (e.g., technical-demo; explain-the-screen; board-instruction), may have mediated students' movement towards instrumental genesis when working on other technology-active problems with Desmos. McMullen et al. (2015) describe this process, where rather than simply providing an answer the teacher encourages the student(s) to use their DT to investigate mathematical ideas, as a *guide-to-investigate* orchestration. It seems that the conscious effort to incorporate dynamic, DT-active representations of functions into these courses, using multiple platforms, supports the crucial role of such representations in epistemic mediation (McMullen et al., 2015; Stacey, 2003). Further, the programs written using GeoGebra have the potential to enable students to move in a dynamic manner from interval-based thinking about AROC of functions to a local perspective. This is achieved by reducing the size of the interval employed for average rate of change from a macro to a micro level, allowing for reflection on the implications of the limiting process as the interval dynamically tends to zero size.

The second study, incorporating the process of using mobile technology to institute two-way, instant, direct, dynamic communication between lecturer and students provides further evidence of the individual *guide-to-investigate* proposed here. In addition, a second individual orchestration we describe as *guide-to-maths-technique* is suggested, where teacher orchestration leads students to make epistemological links between the digital technology techniques and the by-hand techniques. This may be accomplished by enhancing representational versatility (Thomas, 2008) for a technique such as completing the square.

Of course, innovation of practice alone is insufficient to recommend its implementation in teaching. So we must ask what was the effect of the practice outlined above on students? We do not provide a full discussion of the effects on learning here. However, in the first study we saw above that the level of engagement with DT in student learning activity increased, they often responded positively in the survey to its use, and they displayed positive attitudes to DT use in the attitude scale given at the end of the course. In the second study it appears that the innovative use of KakaoTalk as a communication device, along with the individual and whole class instrumental orchestrations by the lecturer described above, may have facilitated communication that contributed to a more positive student perception of calculator use in their learning. For example, the student evaluation, which followed the teaching in the second study, produced comments such as "The application of graphing calculator could be easily understood.", "Conducting classes to engage students was very good", "Using the graphing calculator was a lot of help to conceptually understand and easy to understand mathematics." and "It was nice to learn new mathematics using a graphing calculator". Further, the students were given an attitude scale prior to the course, and again at the end. For around a third of the items there was a significant positive shift in attitude towards DT use from the students. For example, Table 2 shows the mean responses to some of the questions.

Thus we propose that the use of DT during class time, along with visual communication via a smartphone app, may have encouraged student engagement, with the potential for improved conceptual understanding.

Overall, we feel that the kinds of DT-active approaches detailed in this chapter have the potential to assist in smoothing any discontinuity in technology use between school and university. We suggest there are several key aspects to

Table 2 Some of the attitude scale items showing significant positive movement

Scale item	Pre-test mean (N = 214)	Post-test mean (N = 148)	*p*-value
When I use a calculator my learning improves	2.88	3.20	<0.005
More creative problem solving methods come to mind if I use a graphing calculator	2.70	3.14	<0.0005
Studying is more fun when you use tools such as the graphing calculator	3.10	3.40	<0.005
Using a graphing calculator to solve problems makes them more memorable for longer	2.70	3.14	<0.0005

successfully implementing such a scheme if we wish to promote active student engagement and investigation with DT. These include the role of instrumental orchestrations in facilitating instrumental genesis, and the level of teacher-privileging provided for particular technologies. We agree with Carreira, Clark-Wilson, Faggiano and Montone (Chap. From Acorns to Oak Trees: Charting Innovation Within Technology in Mathematics Education) that restructuring lessons with DT, by redefining objectives and tasks "presents new challenges for teachers as they begin to create mathematical tasks that use the affordances of such software environments to produce productive activity for their students". Hence, a crucial factor in any consideration of teaching mathematics with DT is the level of pedagogical technology knowledge (PTK) of the lecturer (Thomas & Hong, 2005; Hong & Thomas, 2006). The PTK lens includes the orientations of teachers and lecturers, a vital factor that is missing in the TPACK framework (Mishra & Koehler, 2006) that is exemplified in the work of Tabach and Slutzky (Chap. Studying The Practice of High SchoolMathematics Teachers in a Single Computer Setting). These orientations form part of Schoenfeld's (2010) resources, orientations and goals model, and without them it is not really possible to gain full insight into the goals of teachers and lecturers (see e.g., Thomas & Palmer, 2013). Their influence on mathematician lecturers, recently described by Schoenfeld, Thomas, and Barton (2016), illustrates why it is important to include them in research. The manner in which PTK may be enhanced for individual lecturers in the digital era (Clark-Wilson, Sinclair, & Robutti, 2013) should therefore continue to be a focus of on-going research and professional development.

References

Artigue, M. (2002). Learning mathematics in a CAS environment: The genesis of a reflection about instrumentalisation and the dialectics between technical and conceptual work. *International Journal of Computers for Mathematical Learning, 7,* 245–274.

Artigue, M. (2010). *The school university interface*. PowerPoint of a seminar given at Auckland University, April 2010.

Bookman, J., & Friedman, C. P. (1999). The evaluation of Project Calc at Duke University 1989–1994. In B. Gold, S. Keith, & W. Marion (Eds.), *Assessment practices in undergraduate mathematics* (pp. 253–256). Washington DC: The Mathematical Association of America.

Blyth, B., & Labovic, A. (2009). Using Maple to implement eLearning integrated with computer aided assessment. *International Journal of Mathematical Education in Science and Technology, 40*(7), 975–988.

Bressoud, D., Mesa, V., & Rasmussen, C. (Eds.). (2015). *Insights and recommendations from the MAA national study of College Calculus*. Washington DC: MAA Press.

Burrill, G., Allison, J., Breaux, G., Kastberg, S., Leatham, K., & Sanchez, W. (2002). *Handheld graphing technology in secondary mathematics: Research findings and implications for classroom practice*, Michigan State University: Texas Instruments. Available from: http://education.ti.com/sites/UK/downloads/pdf/References/Done/Burrill,G.%2520%282002%29.pdf

Cheung, A., & Slavin, R. E. (2011). *The effectiveness of education technology for enhancing reading achievement: A meta-analysis*. Retrieved on February, 8th, 2013, from http://www.bestevidence.org/reading/tech/tech.html

Clark-Wilson, A., Sinclair, N., & Robutti, O. (Eds.). (2013). *The mathematics teacher in the digital era*. Dordrecht: Springer.

Cobb, P., Confrey, J., diSessa, A., Lehrer, R., & Schauble, L. (2003). Design experiments in educational research. *Educational Researcher, 32*(1), 9–13.

Coffland, A., & Xie, Y. (2015). The 21st century mathematics curriculum: A Technology enhanced experience. In X. Ge et al. (Eds.), *Emerging technologies for STEAM education, Educational communications and technology: Issues and innovations* (pp. 311–329). Zurich: Springer. doi:10.1007/978-3-319-02573-5_17

Drijvers, P., Doorman, M., Boon, P., Reed, H., & Gravemeijer, K. (2010). The teacher and the tool: Instrumental orchestrations in the technology-rich mathematics classroom. *Educational Studies in Mathematics, 75*(2), 213–234.

Drijvers, P., Tacoma, S., Besamusca, A., Doorman, M., & Boon, P. (2013). Digital resources inviting changes in mid-adopting teachers' practices and orchestrations. *ZDM: The International Journal on Mathematics Education, 45*(7), 987–1001.

Ellington, A. J. (2003). A meta-analysis of the effects of calculators on students' achievement and attitude levels in precollege mathematics classes. *Journal for Research in Mathematics Education, 34*(5), 433–463.

Fernández-Plaza, J. A., Rico, L., & Ruiz-Hidalgo, J. F. (2013). Concept of finite limit of a function at a point: Meanings and specific terms. *International Journal of Mathematical Education in Science & Technology, 44*(5), 699–710.

Genossar, S.; Botzer, G., & Yerushalmy, M. (2008). Learning with mobile technology: A case study with students in mathematics education. *Proceedings of the CHAIS conference, Open University*. Available from: http://telem-pub.openu.ac.il/users/chais/2008/evening/3_2.pdf

Goos, M., Galbraith, P., Renshaw, P., & Geiger, V. (2003). Perspectives on technology mediated learning in secondary school mathematics classrooms. *Journal of Mathematical Behavior, 22* (1), 73–89.

Graham, A. T., & Thomas, M. O. J. (2000). Building a versatile understanding of algebraic variables with a graphic calculator. *Educational studies in mathematics, 41*(3), 265–282.

Hong, Y. Y., & Thomas, M. O. J. (2006). Factors influencing teacher integration of graphic calculators in teaching. *Proceedings of the 11th Asian technology conference in mathematics* (pp. 234–243). Hong Kong.

Hong, Y. Y., & Thomas, M. O. J. (2015). Graphical construction of a local perspective on differentiation and integration. *Mathematics Education Research Journal, 27,* 183–200. doi:10.1007/s13394-014-0135-6

Hoyles, C., Kalas, I., Trouche, L., Hivon, L., Noss, R., & Wilensky, U. (2010). Connectivity and virtual networks for learning. In C. Hoyles & J.-B. Lagrange (Eds.), *Mathematical education and digital technologies: Rethinking the terrain* (pp. 439–462). New York: Springer.

Jaworski, B., Robinson, C., Matthews, J., & Croft, A. C. (2012). Issues in teaching mathematics to engineering students to promote conceptual understanding: A study of the use of GeoGebra and inquiry-based tasks. *The International Journal for Technology in Mathematics Education, 19*(4), 147–152.

Kendal, M., & Stacey, K. (2001). The impact of teacher privileging on learning differentiation. *International Journal of Computers for Mathematical Learning, 6*(2), 143–165.

Kynigos, C. (2016). Constructionist mathematics with institutionalized infrastructures: The case of Dimitris and his students. In E. Faggiano, F. Ferrara, & A. Montone (Eds.), *Innovation and technology enhancing mathematics education perspectives in the digital era.* Dordrecht: Springer.

Lagrange, J.-B., Artigue, M., Laborde, C., & Trouche, L. (2003). Technology and mathematics education: A multidimensional study of the evolution of research and innovation. In A. J. Bishop, M. A. Clements, C. Keitel, J. Kilpatrick, & F. K. S. Leung (Eds.), *Second international handbook of mathematics education* (pp. 239–271). Dordrecht: Kluwer Academic Publishers.

Larsen, S., Glover, E., & Melhuish, K. (2015). Beyond good teaching. In D. Bressoud, V. Mesa, & C. Rasmussen (Eds.), *Insights and recommendations from the MAA national study of college calculus* (pp. 93–106). Washington DC: MAA Press.

Li, Q., & Ma, X. (2010). A meta-analysis of the effects of computer technology on school students' mathematics learning. *Educational Psychology Review, 22,* 215–243.

Lin, C. T., & Thomas, M. O. J. (2011). Student understanding of Riemann integration: The role of the dynamic software GeoGebra. In J. Hannah, M. O. J. Thomas, & L. Sheryn (Eds.), *Proceedings of Volcanic Delta 2011, The Eighth Southern Hemisphere conference on teaching and learning undergraduate mathematics and statistics, Rotorua, New Zealand, 27 November–2 December 2011* (pp. 216–227). Auckland, New Zealand: The University of Auckland and the University of Canterbury.

McMullen, S., Oates, G., & Thomas, M. O. J. (2015). An integrated technology course at university: Orchestration and mediation. In K. Beswick, T. Muir, & J. Wells (Eds.), *Proceedings of the 39th Conference of the International Group for the Psychology of Mathematics Education* (Vol. 1, pp. 249–257). Hobart, Australia: PME.

Meel, D. E. (1998). Honors students' calculus understandings: Comparing calculus and Mathematica and traditional calculus students. *CBMS Issues in Mathematics Education, Providence, Rhode Island: American Mathematical Society, 7,* 163–215.

Mishra, P., & Koehler, M. (2006). Technological pedagogical content knowledge: A framework for teacher knowledge. *Teachers college record, 108*(6), 1017–1054. doi: 10.1111/j.1467-9620.2006.00684.x

Mousley, J., Lambdin, D., & Koc, Y. (2003). Mathematics teacher education and technology. In A. J. Bishop, M. A. Clements, C. Keitel, J. Kilpatrick, & F. K. S. Leung (Eds.), *Second international handbook of mathematics education* (pp. 395–432). Dordrecht: Kluwer Academic Publishers.

Ng, W. L. (2011). Using an advanced graphing calculator in the teaching and learning of calculus. *International Journal of Mathematical Education in Science & Technology, 42*(7), 925–938.

Norris, C., Soloway, E., Tan, C., & Looi, C. (2013). Inquiry pedagogy and smartphones: Enabling a change in school culture. *Educational Technology, 53*(4), 33–40.

Oates, G. N. (2011). Sustaining integrated technology in undergraduate mathematics. *International Journal of Mathematical Education in Science and Technology, 42*(6), 709–721.

Oates, G., Sheryn, L., & Thomas, M. O. J. (2014). Technology-active student engagement in an undergraduate mathematics course. In P. Liljedahl, S. Oesterle, C. Nicol, & D. Allan (Eds.), *Proceedings of the 38th conference of the international group for the psychology of mathematics education* (Vol. 4, pp. 330–337). Vancouver, Canada: PME.

O'Malley, P., Jenkins, S., Wesley, B., Donehower, C., Rabuck, D., & Lewis, M. E. B. (2013). Effectiveness of using iPads to build math fluency. *Paper presented at the 2013 Council for Exceptional Children Annual Meeting in San Antonio, Texas.* Retrieved 1/9/2016 from http://files.eric.ed.gov/fulltext/ED541158.pdf

Park, K., & Travers, K. J. (1996). A comparative study of a computer-based and a standard college first-year calculus course. *CBMS Issues in Mathematics Education, Providence, Rhode Island: American Mathematical Society, 6,* 155–176.

Paterson, J., Thomas, M. O. J., & Taylor, S. (2011). Decisions, decisions, decisions: What determines the path taken in lectures? *International Journal of Mathematical Education in Science and Technology, 42*(7), 985–996.

Pierce, R., & Stacey, K. (2011). Using dynamic geometry to bring the real world into the classroom. In L. Bu, & R. Schoen (Eds.), *Model-centered learning modeling and simulations for learning and instruction* (Vol. 6) (pp 41–55). Sense Publishers.

Pierce, R., Stacey, K., & Wander, R. (2010). Examining the didactic contract when handheld technology is permitted in the mathematics classroom. *ZDM International Journal of Mathematics Education, 42,* 683–695. doi:10.1007/s11858-010-0271-8

Rasmussen, C., & Wawro, M. (2016, in Press). *Post-calculus research in undergraduate mathematics education.*

Schoenfeld, A. H. (2010). *How we think: A theory of goal-oriented decision making and its educational applications.* New York: Routledge.

Schoenfeld, A., Thomas, M. O. J., & Barton, B. (2016). On understanding and improving the teaching of university mathematics. *International Journal of STEM Education, 3*(4). doi:10.1186/s40594-016-0038-z

Schwingendorf, K. E. (1999). Assessing the effectiveness of innovative educational reform efforts. In B. Gold, S. Keith, & W. Marion (Eds.), *Assessment practices in undergraduate mathematics* (pp. 249–252). Washington DC: The Mathematical Association of America.

Selinski, N. E., & Milbourne, H. (2015). The institutional context. In D. Bressoud, V. Mesa, & C. Rasmussen (Eds.), *Insights and recommendations from the MAA national study of college calculus* (pp. 31–44). Washington DC: MAA Press.

Stacey, K. (2003). Using computer algebra systems in secondary school mathematics: Issues of curriculum, assessment and teaching. In S-C. Chu, W-C. Yang, T. de Alwis & M-G. Lee (Eds.), *Technology connecting mathematics, Proceedings of the 8th Asian technology conference in mathematics*, Taiwan R.O.C: ATCM.

Stewart, S., Thomas, M. O. J., & Hannah, J. (2005). Towards student instrumentation of computer-based algebra systems in university courses. *International Journal of Mathematical Education in Science and Technology, 36*(7), 741–750.

Thomas, M. O. J. (2008). Developing versatility in mathematical thinking. *Mediterranean Journal for Research in Mathematics Education, 7*(2), 67–87.

Thomas, M. O. J., De Freitas Druck, I., Huillet, D., Ju, M. -K., Nardi, E., Rasmussen, C., & Xie, J. (2015). Key mathematical concepts in the transition from secondary school to university. *Proceedings of The 12th international congress on mathematical education (ICME-12) Survey Team 4* (pp. 265–284). Seoul, Korea.

Thomas, M. O. J., & Holton, D. (2003). Technology as a tool for teaching undergraduate mathematics. In A. J. Bishop, M. A. Clements, C. Keitel, J. Kilpatrick, & F. K. S. Leung (Eds.), *Second international handbook of mathematics education* (Vol. 1, pp. 347–390). Dordrecht: Kluwer.

Thomas, M. O. J., & Hong, Y. Y. (2005). Teacher factors in integration of graphic calculators into mathematics learning. In H. L. Chick & J. L. Vincent (Eds.), *Proceedings of the 29th conference of the international group for the psychology of mathematics education* (Vol. 4, pp. 257–264). Melbourne: University of Melbourne.

Thomas, M. O. J., Monaghan, J., & Pierce, R. (2004). Computer algebra systems and algebra: Curriculum, assessment, teaching, and learning. In K. Stacey, H. Chick, & M. Kendal (Eds.), *The teaching and learning of algebra: The 12th ICMI study* (pp. 155–186). Norwood, MA: Kluwer Academic Publishers.

Thomas, M. O. J., & Palmer, J. (2013). Teaching with digital technology: Obstacles and opportunities. In A. Clark-Wilson, N. Sinclair, & O. Robutti (Eds.), *The mathematics teacher in the digital era* (pp. 71–89). Dordrecht: Springer.

Tobin, P., & Weiss, V. (2011). Teaching differential equations in undergraduate mathematics: Technology issues for service courses. In J. Hannah, M. O. J. Thomas, & L. Sheryn (Eds.), *Proceedings of Volcanic Delta 2011, The Eighth Southern Hemisphere Conference on Teaching and Learning Undergraduate Mathematics and Statistics, Rotorua, New Zealand, 27 November–2 December 2011* (pp. 375–385). Auckland, New Zealand: The University of Auckland and the University of Canterbury.

Trouche, L. (2004). Managing the complexity of human/machine interactions in computerized learning environments: Guiding students' command process through instrumental orchestrations. *Educational Studies in Mathematics, 9,* 281–307.

Trouche, L., & Drijvers, P. (2010). Handheld technology for mathematics education: Flashback into the future. *ZDM, The International Journal on Mathematics Education, 42*(7), 667–681. doi:10.1007/s11858-010-0269-2

Vandebrouck, F. (2011). Perspectives et domaines de travail pour l'étude des fonctions. *Annales de Didactiques et de Sciences Cognitives, 16,* 149–185.

White, T., Booker, A., Carter Ching, C., & Martin, L. (2011). Integrating digital and mathematical practices across contexts: A manifesto for mobile learning. *International Journal of Learning and Media, 3*(3), 7–13. doi:10.1162/ijlm_a_00076

The Duo "Pascaline and e-Pascaline": An Example of Using Material and Digital Artefacts at Primary School

Michela Maschietto and Sophie Soury-Lavergne

Abstract The paper presents the design and the analysis of teaching experiments at primary school concerning the introduction and use of a "duo of artefacts", constituted by the pascaline i.e., the arithmetical machine Zero+1, and its digital version e-pascaline. The idea of 'duo of artefacts' represents the innovative component of this research work, because the e-pascaline is constructed in a complementary way with respect to the pascaline. The duo of artefacts is proposed to support student's conceptualization processes of numbers as sign of a quantity, number sequences and recursive addition. Computation and manipulation of base ten notation are two processes that students often consider separately. This duo enables the design of situations that required those two processes to be connected and to consider their effect on each other. With duo of artefacts, technology allows the development of learning environments in which it is possible to study the articulation between material and digital manipulatives for mathematical conceptualization.

1 Connecting Manipulatives and Technological Tools

Research in education is more and more interested in studying the role of manipulatives and their characteristics in teaching and learning, not only mathematics but also sciences. International literature distinguishes between virtual and physical manipulatives. In the recent book "*Perspectives on Teaching and Learning Mathematics with Virtual Manipulatives*", Moyer-Packenham and Bolyard (2016)

M. Maschietto (✉)
Università di Modena e Reggio Emilia, Modena, Italy
e-mail: michela.maschietto@unimore.it

S. Soury-Lavergne
S2HEP Institut Français de l'Education ENS de Lyon, Lyon, France
e-mail: sophie.soury-lavergne@ens-lyon.fr

E. Faggiano et al. (eds.), *Innovation and Technology Enhancing Mathematics Education*, Mathematics Education in the Digital Era 9,
DOI 10.1007/978-3-319-61488-5_7

discuss extensively the definition of virtual[1] manipulatives, starting from the previous definition by Moyer, Bolyard, and Spikell (2002). Their updated definition of a virtual manipulative with respect to other technological tools makes the presence of dynamic representations as an essential feature, and also includes some programmable aspects. Among the five kinds of virtual manipulative environments they have identified, this paper—about the digital part of the duo of artefacts that will be presented later—focuses on "tutorial virtual manipulative environment" and "simulation virtual manipulative environment". But this paper also considers the physical manipulatives and is interested in relationships between virtual (digital) and physical ones. This new attention towards combining physical and digital is rather new, as stated by Bartolini Bussi and Inprasitha (2015) and constitutes an innovation in the field of digital technology for education. Hence, the interest is already shared among recent research works in conceptual learning in mathematics —in particular (Ladel & Kortenkamp, 2015; Soury-Lavergne & Maschietto, 2015a; Sarama & Clements, 2016). The current focus is on the relevance of sensory-motor experiences for mathematical conceptualization (see Chap. Returning to Ordinality in Early Number Sense: Neurological, Technological and Pedagogical Considerations). In science education, such concern also exists. At university level, Olympiou and Zacharia (2012) have carried out teaching experiments using physical manipulative alone, or virtual manipulative alone or the two together. In their overview of the possible combinations of the manipulatives usages, they identify two distinct settings: a "sequential combination" in which the manipulatives are sequentially used, without explicit support between them, and a "blended combination" in which physical and virtual manipulatives are used in reference to each other. The authors carry out their teaching experiments in the latter setting with the aim of analyzing the affordances of the two kinds of manipulatives for students' learning.

We develop our work about connecting manipulatives and digital tools by following the two perspectives of Carreira et al. (in this book) in their study of innovation development. We have designed a technology, which is a duo of material and digital artefacts (Maschietto & Soury-Lavergne, 2013). Moreover, we have designed classroom tasks partly embedded in the digital technology itself and partly in the description of teaching units using the duo of artefacts (Soury-Lavergne & Maschietto, 2015b). We now study the appropriation of the duo and the classroom tasks by the teachers, which is a third perspective about innovation and technology, as in Kynigos's or Tabach and Slutzky's contributions in (Chap. Innovations Through Institutionalized Infrastructures: The Case of Dimitris, His Students and Constructionist Mathematics and Chap. Studying the Practice of High School Mathematics Teachers in a Single Computer Setting).

[1]Sarama and Clements (2016) do not agree with this term. For them, representations on a screen physically exist, they are not 'virtual'.

1.1 From a Learning Perspective

The use of concrete tools implies physical engagement. Many works from different research fields have established its necessity for mathematical conceptualization (Lakoff & Nunez, 2000; Edwards, Radford, & Arzarello, 2009; Kalenine, Pinet, & Gentaz, 2011). The technological tool of the duo (digital and manipulative) must present some evolution in comparison to the physical one in order to help students overcome some of the manipulatives limits. It should offer students a new opportunity to identify the mathematical properties embedded in the artefact behavior. Moreover, digital parts of the duo may evolve toward a more abstract and conventional representation of mathematical objects. In line with Fyfe, McNeil, Son, and Goldstone (2014), there is no opposition between the use of concrete materials and abstract materials in learning, but rather a continuity that helps students build the mathematical concepts in that it gradually eludes the characteristics of concrete material. Also, it contributes to the construction of a “conceptual field” (Vergnaud, 2009), defined by the set of situations in which a concept is carried out:

> It is at the same time a set of situations and a set of concepts tied together. By this, I mean that a concept’s meaning does not come from one situation only but from a variety of situations and that, reciprocally, a situation cannot be analysed with one concept alone, but rather with several concepts, forming systems (Vergnaud, 2009, p. 86).

Therefore, our proposal is to design duos of artefacts, grounded in the example of pascaline and e-pascaline, by associating a concrete manipulative tool to a digital tool to combine the advantages of both types of learning tools and to overcome some of their limits. Mainly, we use the theory of didactical situations (Brousseau, 2002) and especially the concept of feedback in the interaction subject-milieu to design didactical situations including concrete and digital artefacts (Mackrell, Maschietto, & Soury-Lavergne, 2013). Thus, our question is about the characteristics of the duo of artefacts in order to improve the learning experience of students. But, like any learning resources, especially innovative resources, their potential for mathematical learning also depends on the teachers.

1.2 From a Teaching Perspective

The use of information and communication technology by primary school teachers is still not as developed as it could be expected considering the level of equipment and the learning attainments achieved by technology (see Ravenstein and Ladage (2014) for French schools). One of the reasons may be the teachers’ poor understanding of technology as an added value to learn mathematical concepts. They are not convinced of its usefulness. When considering manipulatives, like base-ten kits, sets of cards, coins, dice and so on, the situation is not similar (Moyer-Packenham, Slakind, & Bolyard, 2008). Teachers are aware of the role of manipulation providing physical and perceptual experience as well as solid mathematical conceptualization as regards longs-standing research on education. Teachers actually use such teaching resources, even though the use of concrete manipulatives may also raise difficulties to students, because they embed perceptual and mechanical elements irrelevant to mathematical knowledge or limit knowledge transfer from one situation to another. Thus, a duo of artifacts may be an invitation for teachers to use both tools and especially technology at primary school level. Currently we have very few elements regarding teachers' appropriation and use of complex set of resources, like a duo of artefacts may be (Gueudet, Bueno-Ravel, & Poisard, 2014).

In this study, we use an ergonomic approach developed by Tricot et al. (2003) to evaluate learning environments with technology. We applied it to the duo of artefacts and its appropriation process by teachers. In particular, Tricot and colleagues distinguish three different dimensions to evaluate the efficiency of a learning technology and its potential use by teachers: utility, usability and acceptability. We apply these three dimensions to the duo of artefacts:

- The duo of artefacts is *useful* if it enables students to learn what is intended to. Usefulness characterizes a relevant technology.
- The duo of artefacts is *usable* if students can easily use it and remember how it works, if it is an efficient tool that generates satisfaction and doesn't produce mistakes. Usability characterizes the efficiency of a technology.
- The duo of artefacts is *acceptable* if it matches constraints of the teaching context (like time, space, organization, resources), and also institutional requests and existing practices and if it is compatible with students' motivations, affects, culture and standards. Acceptability conditions the user's decisions to use the environment.

Utility, usability and acceptability can be evaluated empirically, with the involvement of users at every moment of the tool design process. We have organized a study to explicit the point of view of the teachers about the duo of artefacts.

The purpose of this chapter is first to explicit some principles that might be efficient for the design of a duo of artefacts like pascaline and e-pascaline. Then we present a study conducted with teachers and first grade classes to discuss the duo integration into learning and teaching practices. Finally, the ergonomic approach is a suitable theoretical framework to study innovation in the case of teachers' appropriation of a duo of artefacts.

2 The Pascaline and the e-Pascaline for the Learning of Numbers, Place Value and Computation

We have designed a first duo of artefacts composed of a concrete manipulative, the pascaline, and a digital counterpart, the e-pascaline, with the principle that each artefact complements the other one. The pascaline (Fig. 1) is an arithmetic machine developed after the historical machine of the French mathematician Blaise Pascal. It is already used in Italian research on mathematical machines (Maschietto, 2015).

Fig. 1 The pascaline (*left*) displaying number 122 (the three digits above the *red triangles*) and some usage gestures (*right*, Maschietto 2015)

The e-pascaline (Fig. 2) has been developed with the Cabri Elem technology and is a fixture of a collection of e-books.

The duo pascaline and e-pascaline aims at teaching place value and computation; moreover it offers a rich mathematical experience on numbers to the students.

2.1 From the Pascaline to the Design of the e-Pascaline

The pascaline is a simple mechanical machine made up of gears providing a symbolic representation of three digit numbers and adequate for arithmetic operations. Each of the five wheels has ten cogs. The digits from 0 to 9 are stamped on the lower yellow wheels that display units, tens and hundreds from the right to the left. When the units wheel initialized to 0 rotates fully clockwise, the right upper wheel makes the tens wheel rotate in the same direction one tooth forward. This automatic mechanical motion of each lower wheel mediates the idea of packing ten units into one ten, or ten tens into one hundred. Likewise, the jerky motion of the wheel supports the recursive approach to numbers as it rotates one tooth at a time, adding or subtracting 1 according to the rotation clockwise or anticlockwise. It links addition and subtraction as inverse operations.

We have designed a digital machine with some chosen elements of continuities and discontinuities in relation with the physical pascaline (Maschietto & Soury-Lavergne, 2013). The aim was to help the transfer of students’ ideas concerning relevant mathematical meanings and to hinder those irrelevant to mathematical interpretation at primary school level. In the studies about virtual and digital manipulatives (Moyer-Packenham, 2016; Olympiou & Zacharia, 2012), this component of design is not taken into account. Therefore, we have analyzed both students’ schemes of use and their drawings of the pascaline in order to know what physical components and actions of the machine they have clearly identified. We have selected some of these elements to design the e-pascaline. As a result, the e-pascaline looks like the pascaline (Fig. 2) with just some meaningful differences. For instance, with the e-pascaline, it is no longer possible to directly move the upper wheels. Blocking the upper wheels is a means to reinforce the association between a direction of rotation and an operation (the upper wheels turn in the opposite

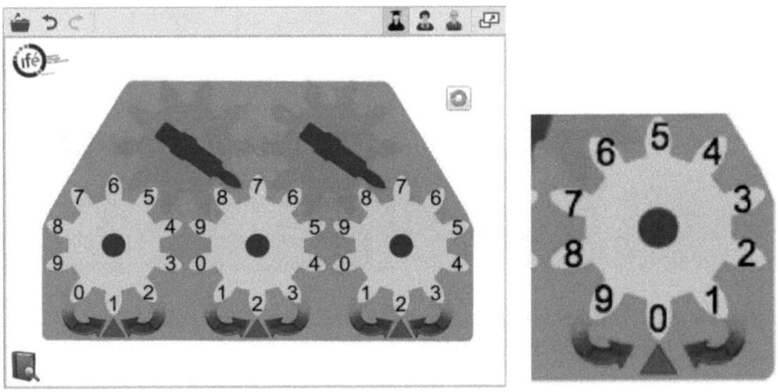

Fig. 2 The e-pascaline (*left*) displaying number 122 and a single wheel with curved arrows for moving the wheel (*right*)

direction due to the principle of gear rotation). Likewise, it is no longer possible to subtract 1 from 0 or to add 1 to 999 (in the pascaline, 000-1 makes 999). The e-pascaline allows only the operations having a result between 0 and 999, which is the full range of numbers displayed by the e-pascaline and the pascaline.

Moreover, the e-pascaline comes with additional components such as action arrows. Indeed, the rotation of the physical pascaline wheels produces sound and haptic feedback each time a tooth revolves. Students use these clicks to control their action on the machine and to perform operations. The e-pascaline makes it explicit by displaying two arrows on each side of the wheels (Fig. 2). These arrows are buttons on which the user clicks to actuate the e-pascaline wheels. In any case, there is always a jerky movement, one by one on a wheel, different from a multi-touch technology (Chap. Returning to Ordinality in Early Number Sense: Neurological, Technological and Pedagogical Considerations). They express a possible action on the digital machine and the direction of the wheel rotation. Other examples of additional components are the reset button (to reset the three wheels to zero) or the "counter of clicks" to display the number of clicks performed by the user since the last reset of the counter (Fig. 3). All these e-pascaline components created on added value and could be used to design problem-solving situations (for instance, writing a number with the minimum of clicks).

The last additional element of the digital part to the duo of artefacts is the collection of e-books offering several possible didactical situations with the e-pascaline.

2.2 *Three Different Levels of Feedback in the Duo of Artefacts*

The didactical situations with the e-pascaline are developed in e-books created with the Cabri Elem technology. From one page to the other, the designers must set the didactical variable values and implement appropriate feedback to cause the

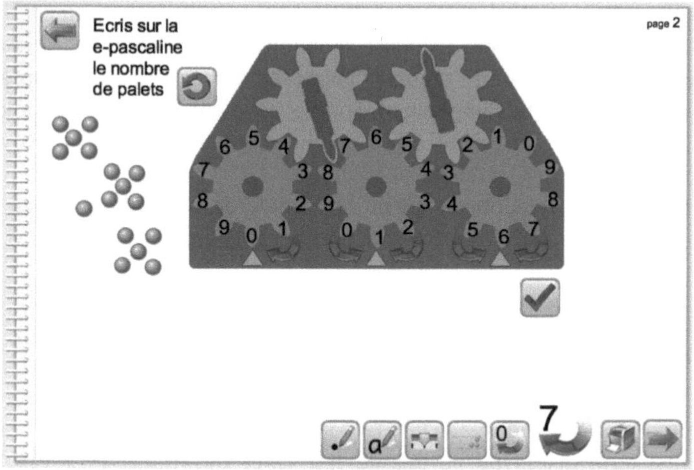

Fig. 3 The e-pascaline in the e-book, with additional components

evolution of the students’ strategies in solving an assigned task (Mackrell et al., 2013). One of the most important design principles at the basis of Cabri Elem authoring environment is direct manipulation (Laborde & Laborde, 2011) which involves both action and feedback on action. In the process of situation design with the authoring environment, we have identified three kinds of feedback.

Direct manipulation feedback is the response of the environment to any students’ action and may be combined to produce the other two types of feedback. An example of direct manipulation feedback is the fact that the rotation of the e-pascaline wheels is displayed continuously when the students click on the action arrow, not only the initial and final state of the wheel. The implementation of a direct manipulation feedback resides mainly in choosing what elements are displayed or hidden, the successive positions of these elements and their dynamic update (the counter of click is automatically updated when clicking on a wheel).

Strategy feedback aims at supporting students in their solving strategy of a given task. It is a response to the mathematical value of the students’ strategy (Brousseau, 2002). To implement strategy feedback, the designers need to identify (i) configurations that are typical of a strategy and induce a diagnosis of this strategy and (ii) new objects or actions that can provide help to the students without changing the nature of the task or giving the answer. Such feedback may consist of help alerts, signs pointing out some contradictory elements in the students’ strategy that call their attention to their current strategy limitations or changes in direct manipulation feedback. Below, we present examples of strategy feedback in the e-pascaline e-books.

Evaluation feedback is related to the completion of the task. Such feedback is necessary for the students to know how successful their strategies are to solve the problem. In the e-pascaline collection of e-books, it is mainly a smiling smiley displayed on the page that indicates success and a sad smiley that indicates errors. Moreover, the successive smileys obtained after each problem remain on the page.

They provide information about the global achievement of the students. It is worth to notice that if the evaluation feedback is automatically displayed, it may happen independently of the students' request. Then, the students can develop a trial and error strategy, seeking the unplanned pop-up of a smiley without looking for a solution. So, in the e-pascaline e-books, the evaluation is given only after an explicit request from the students.

These three levels of feedback appear relevant to design and to analyze didactical situations including each kind of artefact, either physical or digital.

2.3 Adding with the Pascaline and the e-Pascaline

There are two main procedures to add two numbers with the pascaline, both starting from the first term displayed on this device. Once the first term is displayed, the iterative procedure consists in repeating the operation of pushing the units wheel, one tooth at a time clockwise until the number of clicks corresponds to the second term of the sum. For instance, when adding 26 by iteration, the student clicks 26 times on the units wheel. The decomposing procedure consists in pushing each of the three wheels by a number of clicks equal to the corresponding digit of the second term. For instance, when adding 26 by decomposing, the user clicks 6 times on the units wheel and twice on the tens wheel (the order between the wheels does not matter). The iterative procedure is based on the quantity represented by the number while the decomposing procedure is based on the meaning of each of the digits in place value notation. Hence, the evolution of students' procedures from iteration to decomposition corresponds to the transition from a procedure based on the quantity represented by the number (adding by counting one by one) to a procedure based on place value notation. It indicates an evolution of the mathematical meanings associated to place value notation and their possible use for performing operation.

The use of the pascaline provides students with different kinds of feedback. There is an asymmetric direct manipulation feedback when adding two numbers. The feedback is not the same for the two terms of the addition. When the students add two numbers, the pascaline continuously displays the first term and never does for the second one. For the second term, the feedback is reduced to the clicks produced by the moving wheels and the boosted haptic feedback when two or three of the lower wheels turn simultaneously. Using the physical pascaline also enables the students to realize that the two procedures are not equivalent. Therefore, the pascaline provides a kind of strategy feedback. For instance, adding a large number like 100 requires 100 clicks on the units wheel and only one click on the hundreds wheel. Students may be conscious of this difference but the physical machine will not require the evolution from one to the other procedure. And finally, there is no evaluation feedback with the pascaline. This last level of feedback relies on the intervention of a human agent, mainly a teacher.

The goal of the addition e-book is the crucial and tricky passage from the iterative procedure to the decomposing procedure. It corresponds to the evolution from a number representing a quantity, unit by unit, to the writing of number in terms of hundreds, tens and units. Most six-year-old students apply the iterative procedure even with large numbers (Soury-Lavergne & Maschietto, 2015b). The e-book consists of three pages displaying the same structure and components (Fig. 4). The differences from one page to another concern the size of the proposed numbers for addition (up to 30 in pages 1 and 2, up to 69 in page 3) and the type of feedback given by the e-pascaline in response to the students’ procedures. We have implemented feedback to compel students’ procedures to evolve from iteration to decomposition. We used the possibility of hiding the action arrows on the units wheel to compel the students to consider and use another wheel, the tens one. It is possible and efficient because the iteration procedure requires only addition on the units wheels, although the decomposing procedure requires the use of the units wheel and the tens wheel as soon as the second term is a two-digit number. In the first page of the e-book, all procedures are feasible. It supports appropriation of the situation by the students. In the next two pages, the units wheel can only be used a number of times equal to the sum of the unit digits of the two terms. For example, to add 18 + 13 (Fig. 4), the user can only click 8 + 3 times on the units wheel before the addition arrow disappears. The iteration procedure, which needs 13 clicks on the units wheel, is no longer possible. In such a way, students have to look for another strategy to perform the addition. The fact that the action arrow is concealed corresponds to a strategy feedback. It occurs in response to the iterative procedure and shows to the students that iteration on the units wheel is no longer possible, yet does not give the appropriate procedure. Such a feedback is not possible with the physical pascaline. The possibility to design different kinds of feedback contributes to the added value of the e-pascaline to the duo of artefacts.

The differences between the situations with the pascaline and the situation with the e-pascaline are sufficiently clear-cut to support students’ grasp and change of procedures and therefore conceptualization of place value notation. Situations including the pascaline and the e-pascaline involve concrete manipulatives, iconic and symbolic representations with strong links between the three aspects. The duo

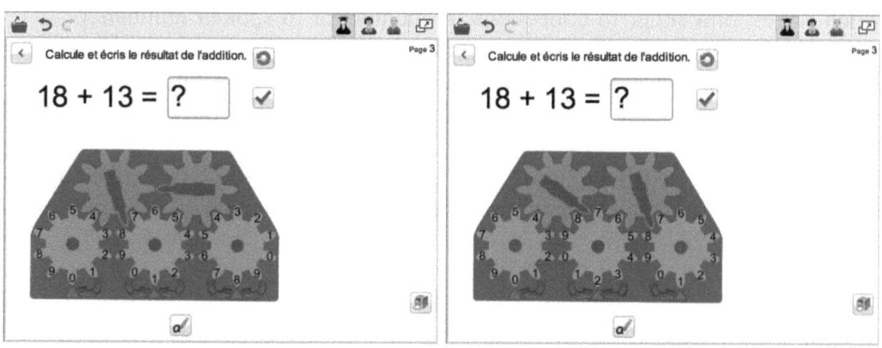

Fig. 4 On the *left*, the first term 18 is written; the e-pascaline is waiting for the second term. On the *right*, the second term is added, by using the units wheel; after three clicks, the adding unit arrow disappears

of artefacts makes possible to create situations and problems that link physical manipulations to different kinds of representation and particularly symbolic representation.

3 Learning Setting with a Duo of Artefacts

The first teaching experiments with the duo of artefacts have been conducted within two French projects with the aim to produce teaching resources based on the duo of artefacts (Soury-Lavergne & Maschietto, 2013). The teachers involved in these projects have started with the e-pascaline, some e-books and some teaching resources concerning the physical pascaline alone [some of them are included in Maschietto and Savioli (2014)]. They have elaborated a teaching scenario, and tested it during two years in different classes. A scenario has been released on one of the projects website[2] to be accessible to any other teachers (Fig. 5). This point is essential in combining the two kinds of artifacts, as Olympiou and Zacharia (2012) claim the need of "*having certain knowledge and skills*" for teachers involved in teaching experiments with them.

The teaching scenario is organized in four units, beginning with the discovery of the physical machine mechanism and ending with a problem solving situation (i.e., the minimum number of clicks for writing given numbers) involving the e-pascaline.

During the first teaching unit, the students discover the physical machine and its mechanical behavior, without being told that it is a mathematical machine. They verbally describe the machine and draw it on a paper. They formulate hypothesis about its possible usages.

In the second teaching unit, they learn to write 2-digits numbers with the physical pascaline, either with an iterative procedure starting from 000 and incrementing the units wheel or by directly using each wheels. The e-pascaline is introduced collectively to gather the hypothesis of the students about how to write a number on the pascaline and the role of each wheel. They use the e-pascaline e-book about writing numbers. The task is to write on the e-pascaline numbers which correspond either to collections of counters or to spoken numbers.

In the third teaching unit, additions and subtractions are introduced with the physical machine and with the e-pascaline. The physical pascaline allows the two types of procedures. Yet the e-pascaline should block the iterative procedure and require the decomposing procedure.

In the fourth teaching unit, the e-pascaline is used to create a new problem-solving situation, which consists in writing a number on the e-pascaline with a minimum of clicks.

---

[2]http://ife.ens-lyon.fr/sciences21/ressources/sequences-et-outils/pascaline-CP. Accessed July 15, 2016.

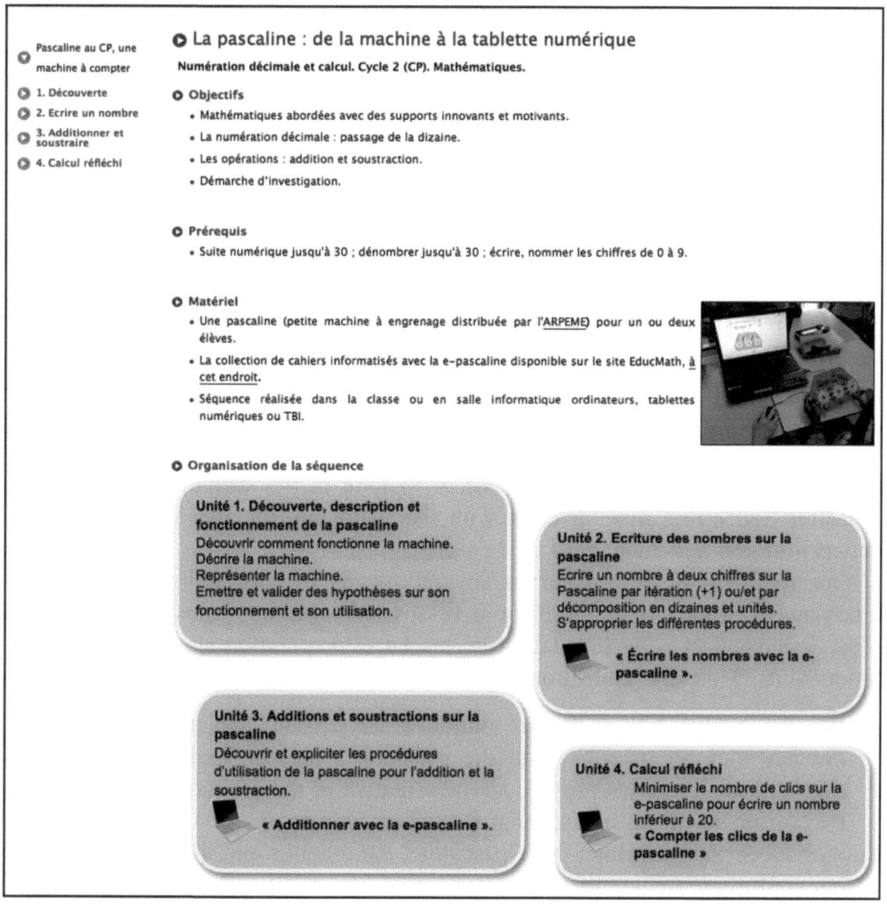

Fig. 5 The teaching scenario on the Plan Science website. The four teaching units use the pascaline or the e-pascaline or both of them

In the experiment presented in this chapter, the resources available to the teachers were the duo of artefacts and the above scenario. We have also provided some online videos and tutorials.[3]

4 Teachers Teaching with a Duo of Artefacts

In this section, we present the setting in which the questions of planning a learning environment with the duo of artefacts and teachers’ appropriation are considered.

[3]http://educmath.ens-lyon.fr/Educmath/recherche/equipes-associees-13-14/mallette/prototype-mallette/page-accueil-de-la-mallette-cp-ce1. Accessed July 15, 2016.

4.1 Experimental Setting in the "Mathematical Package[4]" Project

We have studied the use of the duo of artefacts pascaline and e-pascaline by seven teachers of the first year of French primary school (six-year old students). These teachers had no connection with teachers of the initial research projects and had never seen the pascaline and the e-books nor used them in class before the experiment. They have been recruited by the local educational institution among volunteers, with the constraint of being a pair of teachers in the same school (one of the teachers leaved the experiment, so a teacher was finally single in her school). We assumed that the possibility to easily interact with a colleague was a condition for a successful commitment in the experiment. The educational authority has approved the experimentation, which has also provided positive institutional conditions for the success of the experiment. Moreover, the teachers' extra work has been compensated. Teachers were asked to look at the scenario and to use the pascaline and e-pascaline in class with their students for at least eight sessions between March and June 2014.

Our methodology consisted in several means of data collection. We have interviewed the teachers, asked them to answer some questions, observed some class sessions and collected data, like teachers' preparation sheets and students' productions. This data collection has been planed as follows:

- We (the two researchers that are authors of the present chapter) have organized a one-day kick-off meeting with the teachers, a teachers' educator which was involved in a previous project, a representative of the educational authority and two members of the institutional publisher. During this first meeting, we have introduced the pascaline, the e-pascaline, the e-books and the online resources, but not fully presented them. We have provided an initiation to the pascaline like in Soury-Lavergne and Maschietto (2013). We have also devoted a large part of the meeting to the organization of the experimental setting (documentations, processes of data collections, schedules…).
- During the next twelve weeks the teachers have implemented lessons with the pascaline and the e-pascaline. We have split this period into four parts, which correspond to two weeks of teaching (there was also two holidays weeks during this period). We have planned interviews by phone with each pair of teachers at the end of each of the four periods. We have created a drop box folder for every teacher to collect every file that they could provide: pictures of the class, students' productions, and lesson plans. For each period, we have asked the teachers to report the preparation of the lessons and the implementation of the

[4]The French project was named "Projet Mallette". It was a national project financially supported by the French ministry of Education and the French Institute of Education. It aimed at designing mathematical tools and learning situations for teaching mathematics at the kindergarten levels and grades 1 and 2.

lessons. About the preparation of lessons (before and after the lessons), we asked them to precise: the date and its approximate duration (for the preparation phase and eventually after the lesson), if their work was individual or collective (with colleagues or pedagogical counselor or any other person), and the resources that had been consulted.

- We have organized a final one-day meeting with several aims: to get some final data, but also to give them feedback of the experiment and the opportunity to discuss their expectations and difficulties about the use of the duo of artefacts. A week before this meeting, teachers were asked to send us a message telling two positive aspects of the experiment and two negative aspects. We have organized their responses into different themes and presented them during the last meeting. The final meeting has also be an opportunity for the teacher to share their creations and adaptations of the teaching scenario.

About the lessons with the students, we asked for the following facts: date, number of students attending the lesson, anticipated and actual duration, computers setting (beamer, computers in the classroom, IWB, computer room), pedagogical setting (collective work, group work, individual work), number of adults present during the lesson, title of the e-books used… We also required for a short description of the lesson and its origin (was the lesson proposed in the scenario or did it come from another source). The teachers had to give commentaries about how the lesson occurred, from different points of view: mathematics at stake and students' learning, unexpected events and surprises, new ideas and changes of plan during the lesson, students' comments and expressed opinions… We also asked them to collect students' productions and eventually some other pictures of the class. We completed this methodology by direct observation of two successive lessons in two classes of a same school.

During this experiment, our methodology was not only a way to collect the data we needed for our study, but also it has been a means to provide assistance to the teachers and to support innovation. Indeed, during interviews, teachers had the opportunity to report their difficulties and their success and we could help, acknowledge and encourage them. This wouldn't be the case of teachers working on their own. Nevertheless, the data were meaningful to study the appropriation of a duo of artefacts.

5 The Duo in the Classes

In this section, we discuss a case in which the students are asked to interpret the result of their actions and their consequent feedback on the pascaline and on the e-pascaline in the e-book about addition (see Sect. 2.3). It concerns the fact that the arrow of the units wheel disappears after a certain number of clicks.

5.1 From the Students' Point of View, Resistance to Abandon Iterative Strategy When Adding Numbers with the Pascaline and the e-Pascaline

Like it was proposed in the scenario, in the implemented teaching sequence, students first worked on addition with the pascaline alone and then used the e-book. We have observed the two inadequate strategies that are regularly appearing during the introduction of the pascaline: (i) the two terms are written on two separate wheels and the result is expected to appear on the third wheel; (ii) the addition was done by mental calculation and the result was written on the pascaline. Some students also proposed a third strategy: (iii) the result of the addition is decomposed in a sum of three one-digit numbers that are finally written on each wheel. The first strategy is analogous to the use of a calculator. The users expect the pascaline to be a two arguments function and they transfer the main part of the work to the pascaline. With the second strategy, the users perform the main part of the work and use the pascaline only as a display device. With the third strategy, the users are working with the additive representation of the number. The iteration procedure can only appear once these previous strategies have been invalidated and after teacher's interventions.

In Cleo's class, the iterative procedure appeared after she suggested using the units wheel alone. Once the students have moved to the iterative procedure, the cost of this new strategy and the level of errors were not high enough to make students looking for another strategy. Even with two terms greater than ten and even when the teacher suggested to look for another strategy, students do not change again to use the decomposing strategy. Only one of Cleo's twenty-three students found and used the decomposing procedure. When Cleo's students first used the e-pascaline e-book for addition, she organized it as an individual activity and students still had the possibility to use the physical pascaline together with the e-pascaline. Hence, many of them used the pascaline, on which the iteration procedure was still possible. Then, Cleo compelled the students to perform addition using the e-book.

In Stina's class, after introducing addition with the pascaline, pairs of students used the e-pascaline and the e-book "addition" on laptops. They all used the iterative procedure and were completely unable to progress once the arrow of the units wheel disappeared. During the next lesson, Stina has transformed this incident into a learning opportunity. She asked her students to describe the problem with the e-pascaline and to share ideas regarding solutions to perform additions. She summarized the collective discussion on the whiteboard (Fig. 6). Students came up with different explanations (for instance), among which: the system enables no more

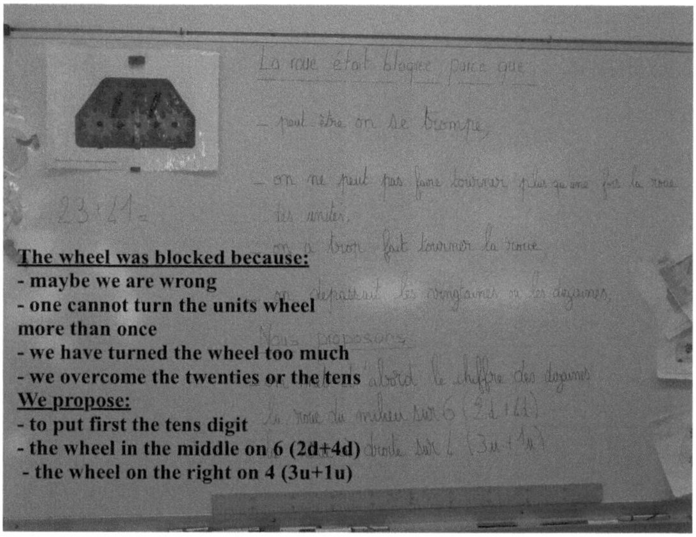

Fig. 6 Students’ explanations for the missing arrow and solutions to compute 23 + 41 with the e-pascaline

than one click, or we must not turn too much the wheel, or we must not make the tens wheel turning (Fig. 7).

Then, Stina asked her students to look for additive decompositions of numbers. They worked in small groups and had to write 23 and 41 in different ways, using the physical pascaline to check their solutions. For number 23, students suggested the following decompositions:

20 + 3; 13 + 10; 10 + 10 + 3; 10 + 5 + 5 + 3; 10 + 10 + 2 + 1; 5 + 5 + 5 + 5 + 2 + 1 and even 11 + 12 and 14 + 9.

During the next lesson, when we observed her students with the e-pascaline, most of them were able to use the two procedures and to compare their efficiency.

The analysis shows that the iterative procedure is the main first grade students’ strategy and that there is a strong resistance of the students to change their procedure. As anticipated, the physical pascaline is not providing a *milieu* that makes the iterative procedure inadequate, even if this procedure may be considered as long and uncertain. On the contrary, the e-pascaline in the e-book offers extra feedback that makes the students aware of the limits of the iterative procedure. But without teachers’ intervention, there is no evolution from the iterative to the decomposing procedure. Moreover, the strategy feedback of the e-pascaline, which consists in hiding the action arrow on the units wheel, is considered as a bug by the students.

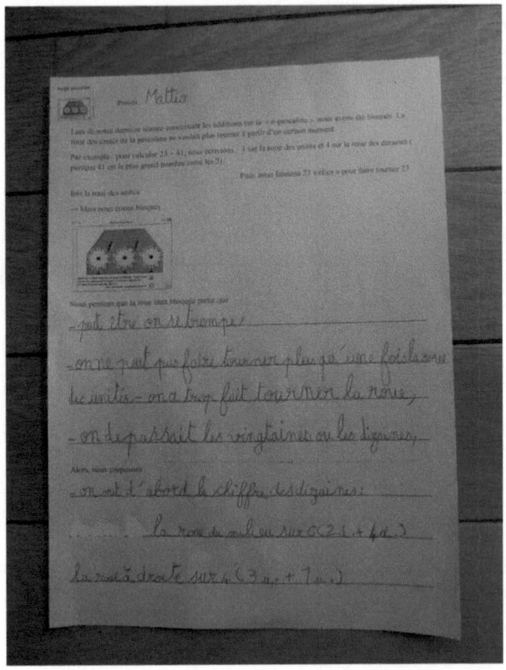

Fig. 7 One student's sheet explaining why the units wheel could be blocked

5.2 *From the Teachers' Point of View*

5.2.1 A Difficulty in Interpreting a Strategy Feedback

From the point of view of the teachers, the disappearing arrow in the e-book about addition was also an unexpected event that occurred during the lessons. Teachers were not aware of this feedback of the e-pascaline although it was explained in commentaries for teachers in the e-book for addition and also demonstrated in a tutorial video. Each of the seven teachers involved in the experiment has been challenged by the phenomena during the lesson with the students. None of them came up with an adequate explanation:

> "So, I don't know if we have done well or not, with Nelly, we have faced the same problem, I don't know what happened but we have quickly been confronted with the problem that the little purple arrow, on the right, went away! So in almost every group, at a certain moment they [the students] were blocked because they couldn't push this right purple arrow. Well so we stopped!" (Stina, period 1).

> "It was mainly the capable students, they have faced a problem. Well well, how to say, if they were exceeding the number, for instance if they want to add 2 tens and being to fast they added 3, there is an arrow that disappears. Then they couldn't move back, well that bothered them. [...] We [the teachers] have been aware of that, with the students, not before." (Lila, period 3).

> "Well a last little thing, just with the e-pascaline e-book. I did a video. It's a pity that… For instance, it was 17 plus 8. A student succeed, he has displayed 17. Then the student turns 8 times the units wheel. But after, the purple arrow, the one to move forward the units wheel, it disappears. So, one cannot turn the units wheel more than 8 times. […] I wonder if it has been done purposely" (Cleo, period 3).

Teachers thought it was a bug of the software or wonder if it has been done on purpose. But after the first surprise and a discussion during the interviews, they understood the reason of the disappearing arrow and turned the phenomena into a pedagogical opportunity.

Stina and Nelly have planned a new lesson. They have transformed the situation into a new problem, which consists to understand why the e-book stops the calculation: "*we are going to see why we are stopped*" (Stina, reporting her interaction with her class in period 2). The students have proposed several explanations (Fig. 7). Then the teacher has clarified the two procedures, showing that even if the arrow disappears, the calculation may be completed.

Another teacher, Cleo, has questioned the feedback relevance when the second addend is lower than 10. Indeed, the students may not stop incrementing the units wheel, because they have controlled the process and know that they have finished with the calculation. They may stop because they cannot go on any more and still they produce a correct answer. It happens when the second addend is lower than ten. This situation may happened because the additions are randomly generated by the system within a set of constraints which includes checking the fact that the sum of the two units digits of the two addends is over 10 but not checking that the second addend is over 10.

We must precise that displaying or hiding the action arrows is a feedback used in every e-book with the e-pascaline. It prevents any calculation that would result in a number outside the range from 000 to 999. For instance, when the e-pascaline displays 000, the user cannot perform a subtraction with any of the three wheels, because the mathematical result would be a negative integer incoherent with the number displayed on the pascaline. Teachers and their students have easily noticed these phenomena when they were discovering the e-pascaline: "*At the first clicks, some have reacted, they said Yes but there is an arrow that comes in the other way, yeah it's to go in the other direction* […] *and which disappears when you're at zero. Well they also said to me: the arrow is there starting at 1, at zero it disappears.*" (July, period 2). Thus, it is an already known feedback. It should not have caused such a surprise. The teachers' reaction reveals that the meaning of this feedback was not immediately accessible in the situation. Its value as a strategy feedback was not spontaneously identified.

5.2.2 Instrumental orchestration

Using the duo of artefacts also implies for teachers to find the adequate orchestration between the different resources (Drijvers, Doorman, Boon, Reed, & Gravemeijer, 2010). The teachers have found different ways to organize students'

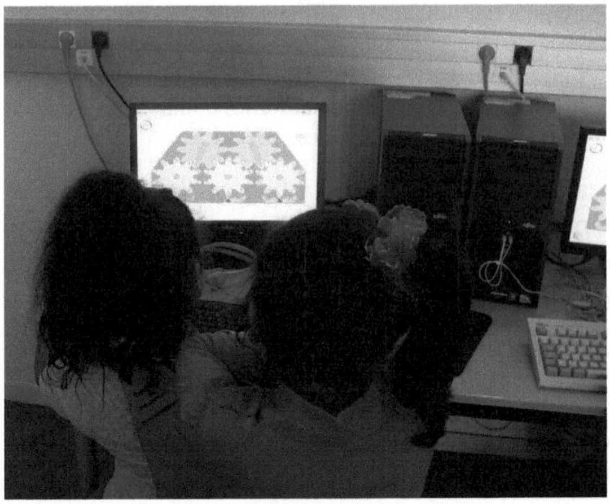

Fig. 8 Two pupils sharing the same computer, one sited and the other standind up (Cleo's class)

Fig. 9 Half of the class is working autonomously in the *middle* of the computer laboratory and the other half of the class is working with the e-pascaline, one pupil per computer (Cleo's class)

access to the duo of artefacts and the e-pascaline on the computers. They have invented configurations of artefacts according to the available equipment (pascaline and e-pascaline, computers, video projector, IWB…). For instance, the following spatial configurations with e-pascaline were proposed:

(1) the e-pascaline in a computer laboratory (Figs. 8 and 9)
(2) the e-pascaline in the usual classroom, on a laptop (Fig. 10)
(3) the e-pascaline available only by video projector (Fig. 11).

Teachers set up different exploration modes. Some of them worked with sub-groups of the class successively (a half of the class or a smaller group of about five students per session). Meanwhile, the other groups were doing something else, in the same room (Fig. 9) or in another room when another adult could supervise them (Fig. 10). These teachers reproduced the same lesson until every student of the class

Fig. 10 Half of the class is working with the e-pascaline in the usual classroom, two pupils per computer. The other half of the class is outside the classroom with another adult (Stina's class)

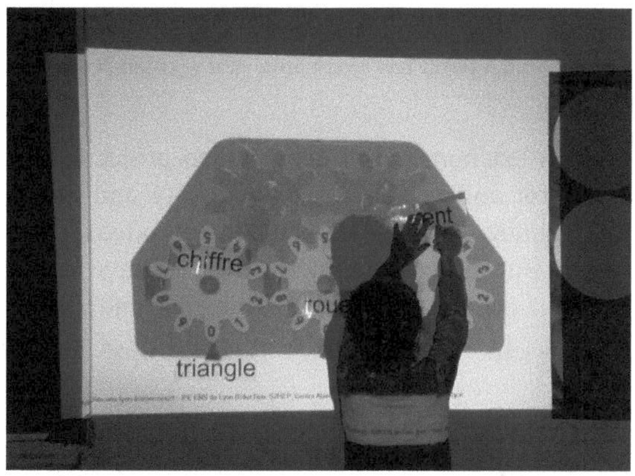

Fig. 11 Labels to legend a video projected image of the pascaline (Cleo's class)

had been through. Some teachers organized work with the duo of artefacts for the whole class at the same time, with the same task or with different tasks (for instance, part of the group with computer, the other with pascaline).

Some teachers organized class work, in their usual classroom, with the physical pascaline and the e-pascaline projected on a screen (Figs. 11 and 12).

Some teachers organized individual access to the pascaline and the e-books and others always make their students work by pair (Fig. 8). One of the teachers has selected only a small sub-group of students to work with them, assuming that the other students were facing too many difficulties to work with the pascaline (Iris, period 3). Five of the seven teachers have used the possibility of video-projecting the e-pascaline to support collective discussion (Fig. 12). Cleo prepared printed labels to legend a projected image of the pascaline (Fig. 11), available in the pack of resources. Rose also brought a Wi-Fi mouse to enable students to easily manipulate the e-pascaline that is video-projected, without leaving their place. This kind of orchestration exploits the added value of the technology in the duo of artefacts.

Fig. 12 Each pupil has a pascaline and a video-projection of the e-pascaline is used (Iris's class)

Two teachers didn't use a video-projector, even if they agree that it would be easier for collective discussion (Nelly and Stina, period 4). Collective discussion was required by the didactical methodology on which the teaching scenario is based. But along the experiment, we have noted some evolutions in the way the teachers organized sharing strategies and discussions about results among their students, giving them the responsibility of moving the wireless mouse.

6 Utility, Usability and Acceptability of a Duo of Artefacts from the Teachers' Point of View

Our study of the duo also includes an evaluation of its utility, usability and acceptability. Teachers have identified positive aspects that can be related either to the utility of the pascaline, or to the utility of the e-pascaline, or to the duo.

About utility to learn mathematics, in a written questionnaire during last meeting of the experiment the teachers declare that: (i) the duo of artefacts provides a complementary conceptual approach of numbers with respect to the number line, the duo helps student to distinguish between units and tens and help to build the notions of complement to 10 and complement to 100; (ii) this duo enables students to visualize numbers and to use different procedures; (iii) students face investigative situations and actual mathematical problems; and (iv) students understand better. They find e-books useful to: organize students' autonomy and differentiation, to arrange exercising and evaluation, to help students. Finally they considered that "*it's new*" compared to any other pedagogical material. They also emphasis the complementarity in the duo: "*a positive aspect is the link with the computers (e-pascaline)*" (Cleo, last meeting) and "*e-books are really complementary and serve to train and evaluate if the notion is learned*" (Rose, last meeting), "*using the e-pascaline in complement with the pascaline enables to reinvest what has been learned in a different way*" (Stina, last meeting), "*and above all to pose problems that, for them, going from the pascaline to the e-pascaline, were much more playful*" (Nelly, last meeting).

They raised also negative points regarding utility. First, the pascaline does not help students with difficulties and some of the e-books are too difficult for the level of schooling (first grade). Second, the ludic and funny aspect of the pascaline may prevent some students to engage into the mathematical situation. Indeed, the ludic dimension of the pascaline is perceived as being both positive and negative. Finally about e-books, even if the teachers had a positive opinion, they identified characteristics that decrease their utility regarding learning. They particularly discussed the fact that students can reload a new problem without having solved the previous one (for instance reload a new couple of numbers to add). Teachers confronted two points of views. On one side, they consider it positively because it enables students to choose a problem adapted to their level of knowledge (which signs that students are aware of what they know and can do, a kind of self evaluation). On the other side, it is negative for the learning because it may give students a way to avoid problems teachers want them to face. In conclusion, teachers have to manage the balance between two possible uses, according to each student's attitude. It requires additional attention and teacher's feedback during the lesson. Different evolutions of the e-book are possible to support the teacher's task. One of them is to keep an explicit trace of every problem launched by the students, with the indication of the solved and unsolved ones.

As for utility, usability for teacher inherits of students' usability. The teachers have noticed that it is easy for the students to engage in the work and to appropriate the pascaline because "*the device is pleasant*" (Lila). Students encountered some difficulties in using the e-books. There has been some mouse and click problems. Some functions were missing, like a zone to write the student's name or a reload button to initialize the e-book without closing it. Each of these problems can be easily solved (the experiment produced several useful indications for updating the e-books). But usability was also related to the teachers' specific tasks. They found the duo of artefacts easy to use because it fits with any textbook and it is easy to manage different students' work pace, by using different e-books simultaneously. But there are also difficulties. The first one is time related. The teachers have implemented up to fifteen lessons for about nine hours. Rose declares that it asked her a lot of time. The other main difficulty is about computers management: management of students' files, management of hardware (some computers launch automatic updates or run out of charge during students' work, school computers are too slow), computers rooms are not big enough to have a sit for every students. It also appears to be difficult for the teachers to be sure that every student has manipulated the pascaline and the e-pascaline by himself. They would also like to have a global view on students' achievement with the e-books, without having to open each student e-book one by one.

The teachers have evaluated positively the acceptability of the duo of artefacts, mainly because the institutional demands are fulfilled. They have declared that the pascaline and the e-pascaline are usable all along the year, not only for a too specific part of the curriculum. It completes the mathematics curriculum about base ten-place value notation, which is a main part of the first grade mathematics curriculum. Using the e-books is adapted to the competencies of the students and the targeted ones. Finally, it responds to the institutional demand about using technology in class.

7 Discussion About Innovation with a Duo of Artefacts

7.1 The Innovation Is About Connecting Physical and Digital Manipulatives

The utility of the duo of artefacts for learning mathematics results form the design of the digital artefact in continuity and discontinuity with regard to the physical one. The e-books implement new feedback that is in continuity with the physical pascaline, to help the student to connect the two artefacts. But it also implements new feedback introducing discontinuity, like the episode of the disappearing arrow shows, to create learning situations. Thanks to the evaluation feedback implemented in the e-books, teachers can also monitor the achievement of the task by the students without interrupting them. It is not possible with the physical pascaline.

In conclusion, feedback and design in the duo of artefacts and in the e-books constitute the innovation. Innovation is also in the complementarity of the physical and digital artefacts in the duo and the existence of a learning scenario that blend them.

The ergonomic analysis has pointed that the teachers are aware of the complementarity between the physical and the digital artefacts of the duo.

7.2 Introducing Digital Technology in Primary School Classes Is Still an Innovation

Until now, introducing digital technology in class is still an innovation when considering primary school and primary school teachers. Does the duo of artefacts promote the introduction of digital technology in class? The ergonomic analysis has showed that the duo of artefacts is considered to be useful by the teachers. Moreover, the acceptability of the duo of artefacts is good and related to the fact that it enables to introduce technology in class, which is required by the institution.

The usability of the duo of artefacts may be improved (the experimentation has provided a list of possible updates and improvements of the e-books). But it is currently already sufficient to enable relevant and efficient classroom practices. Nevertheless, the general agreement about the utility of the duo of artefacts was not really a debate. It may be a bias of our methodology because the teachers are volunteers and certainly want to raise the positive aspect of their involvement in the experiment.

Finally, the duo of artefacts supports the use of the digital technology by the teachers. Every teacher has found her own way to technology and to use the e-pascaline and the e-books in her class. Yet, in the local context of their schools, the teachers have implemented many different didactical configurations. It shows their creativity in the organization of the class. Moreover, these different didactical configurations ask for the orchestration of additional instruments like video projector or IWB. The teachers have found a place for the duo of artefacts pascaline and e-pascaline in their practices.

References

Bartolini Bussi, M. G., & Inprasitha, M. (2015). Theme 3: Aspects that affect whole number learning. In X. Sun, B. Kaur, & J. Novotna (Eds.) (pp. 277–281). Presented at the The Twenty-third ICMI Study: Primary Mathematics Study on Whole Numbers, Macau, China.

Brousseau, G. (2002). *Theory of Didactical Situations in Mathematics*. Springer.

Drijvers, P., Doorman, M., Boon, P., Reed, H., & Gravemeijer, K. (2010). The teacher and the tool: Instrumental orchestrations in the technology-rich mathematics classroom. *Educational Studies in Mathematics, 75*(2), 213–234.

Edwards, L., Radford, L., & Arzarello, F. (Eds.). (2009). Gestures and multimodality in the construction of mathematical meaning. *Educational Studies in Mathematics, 70*(2).

Fyfe, E. F., McNeil, N. M., Son, J. Y., & Goldstone, R. L. (2014). Concreteness fading in mathematics and science instruction: A systematic review. *Educational Psychology Review, 26* (1), 9–25. doi:10.1007/s10648-014-9249-3

Gueudet, G., Bueno-Ravel, L., & Poisard, C. (2014). Teaching mathematics with technologies at Kindergarten: Resources and orchestrations. In A. Clark-Wilson, O. Robutti, & N. Sinclair (Eds.), *The mathematics teacher in the digital era* (Vol. 2, pp. 213–240). Springer.

Kalenine, S., Pinet, L., & Gentaz, E. (2011). The visuo-haptic and haptic exploration of geometrical shapes increases their recognition in preschoolers. *International Journal of Behavioral Development, 35,* 18–26.

Laborde, C., & Laborde, J.-M. (2011). Interactivity in dynamic mathematics environments: What does that mean? In *Integration of Technology into Mathematics Education: past, present and future Proceedings of the Sixteenth Asian Technology Conference in MathematicsTCM*. Bolu, Turkey. Retrieved from http://atcm.mathandtech.org/EP2011/invited_papers/3272011_19113.pdf

Ladel, S., & Kortenkamp, U. (2015). Development of conceptual understanding of place value. In *The twenty-third ICMI study: Primary mathematics study on whole numbers* (p. 323). Macau, China: ICMI.

Lakoff, G., & Nunez, R. (2000). *Where mathematics comes from: How the embodied mind brings mathematics into being*. New York: Basic Books.

Mackrell, K., Maschietto, M., & Soury-Lavergne, S. (2013). The interaction between task design and technology design in creating tasks with Cabri Elem. In C. Margolinas (Ed.), *ICMI study 22 Task design in mathematics education* (pp. 81–90). Royaume-Uni: Oxford.

Maschietto, M. (2015). The arithmetical machine Zero +1 in mathematics laboratory: Instrumental genesis and semiotic mediation. *International Journal of Science and Mathematics Education, 13*(1), 121–144. doi:10.1007/s10763-013-9493-x

Maschietto, M., & Savioli, K. (2014). *Numeri in movimento. Attività per apprendere l'aritmetica con la pascalina.* Collana Artefatti intelligenti. Trento: Erickson.

Maschietto, M., & Soury-Lavergne, S. (2013). Designing a duo of material and digital artifacts: The pascaline and Cabri Elem e-books in primary school mathematics. *ZDM—The International Journal on Mathematics Education, 45*(7), 959–971. doi:10.1007/s11858-013-0533-3

Moyer, P., Bolyard, J. J., & Spikell, M. A. (2002). What are virtual manipulatives? In NCTM (Ed.), *Teaching children mathematics* (pp. 372–377).

Moyer-Packenham, P., & Bolyard, J. J. (2016). Revisiting the definition of a virtual manipulative. In P. Moyer-Packenham (Ed.), *International perspectives on teaching and learning mathematics with virtual manipulatives* (Vol. 7, pp. 3–23). Cham: Springer International Publishing. Retrieved from http://link.springer.com/10.1007/978-3-319-32718-1_1

Moyer-Packenham, P. S. (Ed.). (2016). *International perspectives on teaching and learning mathematics with virtual manipulatives.* Berlin, New York, NY: Springer.

Moyer-Packenham, P., Slakind, G., & Bolyard, J. (2008). Virtual manipulatives used by K-8 teachers for mathematics instruction: Considering mathematical, cognitive, and pedagogical fidelity. *Contemporary Issues in Technology and Teacher Education, 8*(3). Retrieved from http://www.citejournal.org/vol8/iss3/mathematics/article1.cfm

Olympiou, G., & Zacharia, Z. C. (2012). Blending physical and virtual manipulatives: An effort to improve students' conceptual understanding through science laboratory experimentation. *Science Education, 96*(1), 21–47. doi:10.1002/sce.20463

Ravenstein, J., & Ladage, C. (2014). Ordinateurs et Internet à l'école élémentaire française. *Education & Didactique, 8*(3), 9–21.

Sarama, J., & Clements, D. H. (2016). Physical and virtual manipulatives: What is "Concrete"? In P. Moyer-Packenham (Ed.), *International perspectives on teaching and learning mathematics with virtual manipulatives* (Vol. 7, pp. 71–93). Cham: Springer International Publishing. Retrieved from http://link.springer.com/10.1007/978-3-319-32718-1_4

Soury-Lavergne, S., & Maschietto, M. (2013). A la découverte de la «pascaline» pour l'apprentissage de la numération décimale. In C. Ouvrier-Buffet (Ed.), *XXXIXe colloque de la COPIRELEM Faire des mathématiques à l'école: de la formation des enseignants à l'activité de l'élève.* France: Quimper.

Soury-Lavergne, S., & Maschietto, M. (2015a). Articulation of spatial and geometrical knowledge in problem solving with technology at primary school. *ZDM—The International Journal on Mathematics Education, 47*(3), 435–449. doi:10.1007/s11858-015-0694-3

Soury-Lavergne, S., & Maschietto, M. (2015b). Number system and computation with a duo of artefacts: The pascaline and the e-pascaline. In X. Sun, B. Kaur, & J. Novotna (Eds.), *The twenty-third ICMI study: Primary mathematics study on whole numbers* (pp. 371–378). Macau, China: ICMI.

Tricot, A., Plégat-Soutjis, F., Camps, J.-F., Amiel, A., Lutz, G., & Morcillo, A. (2003). Utilité, utilisabilité, acceptabilité. Interpréter les relations entre trois dimensions de l'évaluation des EIAH. In *Proceedings of EIAH* (pp. 391–402). Strasbourg, France.

Vergnaud, G. (2009). The theory of conceptual fields. *Human Development, 52,* 83–94.

What Is Or What Might Be the Benefit of Using Computer Algebra Systems in the Learning and Teaching of Calculus?

Hans-Georg Weigand

Abstract Advantages and disadvantages of the use of Digital Technologies (DT) and especially of Computer Algebra Systems (CAS) in mathematics lessons are worldwide discussed controversially. Many empirical studies show the benefit of the use of DT in classrooms and there are also many useful examples concerning their use. However, despite these inspiring results and the countless ideas, classroom suggestions, lesson plans and research reports, the use of DT—and especially CAS—has not succeeded, as many had expected during the last decades see Hoyles & Lagrange, (2010). The thesis of this article is that we have not been able to convince teachers, lecturers at university and parents of the benefit of CAS in the classrooms in a sufficient way. What are the arguments that justify the use of CAS in the classroom? The article gives examples of a fruitful use of CAS with regard to the generally accepted goals or standards of mathematics education—like fostering students' abilities in problem solving, modelling, proving or communicating—and to the subjects taught in high school. The basis of the argumentation is a competence model which classifies the relation between contents or topics: sequences and limits, functions and equations; representations of DT or CAS: static isolated, static multiple, dynamic isolated and dynamic multiple representations; and classroom activities: calculate, consult, control, communicate and discover.

1 Concerning the Use of DT in Mathematics Lessons

There are many theoretical considerations, empirical investigations and suggestions for the classroom concerning the use of DT and especially CAS in mathematical learning and teaching (Artigue, 2002; Pierce & Stacey, 2004; Guin, Ruthven & Trouche, 2005; Kieran & Drijvers, 2006; Zbiek, 2007; Drijvers & Weigand, 2010;

H.-G. Weigand (✉)
University of Wuerzburg, Würzburg, Germany
e-mail: weigand@mathematik.uni-wuerzburg.de

E. Faggiano et al. (eds.), *Innovation and Technology Enhancing Mathematics Education*, Mathematics Education in the Digital Era 9,
DOI 10.1007/978-3-319-61488-5_8

Weigand, 2013). In recent times, some empirical studies started integrating CAS into regular classroom teaching and covering longer periods of investigation. The *e-CoLab*[1] project in France (Aldon et al., 2008), the one year *TI-Nspire* project (using the non-CAS-version) of the University of Chichester in secondary mathematics classrooms (Clark-Wilson, 2008), the *RITEMATHS*[2] projects in Australia. The *CALIMERO*[3] project (Ingelmann & Bruder, 2007) started with the use of computer algebra 2005 in grade 7 and continued on to the following grades in the next years, and from 2005 to 2013, the "M^3-Project"[4] tested the use of *Symbolic Calculators (SC)* in Bavarian 'Gymnasien' (grammar schools) in Germany with students from grades 10, 11 and 12 (see Weigand, 2008; Weigand & Bichler, 2010b, c).

The main results of these projects and investigations can roughly be summarized as follows: Computer Algebra Systems

- allow a greater variety of strategies in the frame of problem solving processes;
- are a catalyst for individual, partner and group work;
- do not lead to a deficit in paper-and-pencil abilities and mental abilities (if these abilities are regularly supported in the teaching lessons);
- allow more realistic modelling problems in the classroom (but also raise the cognitive level of the understanding of these problems);
- do not automatically lead to changed or modified test and examination problems (compared to paper-and-pencil tests);
- demand and foster advanced argumentation strategies (e.g. if equations are solved by pressing only one button).

However, overall, classroom teaching and learning did not change automatically due to the additional use of a new tool. It needs didactic and methodic considerations, a thorough thinking about the goals of teaching and the possibilities of change as a prerequisite of a gainful change in real classroom teaching (and learning).

2 Visions and Disillusions

The NCTM standards of 1989 (and in the revised version of 2000) have been visionary—concerning the field of mathematics education—by representing a vision for the future of mathematics education. This is especially true for the use of

[1]e-CoLab = Expérimentation Collaborative de Laboratoires mathématiques. See: http://educmath.inrp.fr/Educmath/dossier-parutions/experimentation-collaborative-de-laboratoires-mathematiques. *Accessed 29 May 2016.*

[2]RITEMATHS = The project is about the use of real problems (R) and information technology (IT) to enhance (E) students' commitment to, and achievement in, mathematics (MATHS). http://extranet.edfac.unimelb.edu.au/DSME/RITEMATHS. *Accessed 29 May 2016.*

[3]CALIMERO = Computer Algebra in Mathematics Lessons: Discovering, Calculating, Organizing (translated title).

[4]M^3 = **M**odel Project New **M**edia in **M**athematics Education.

new technologies in mathematics classrooms, expressed in the 'Technology Principle':

> Technology is essential in teaching and learning mathematics; it influences the mathematics that is taught and enhances students' learning. (p. 24)

and:

> Calculators and computers are reshaping the mathematical landscape … Students can learn more mathematics more deeply with the appropriate and responsible use of technology. (p. 25)

The first ICMI study in 1986 "The Influence of Computers and Informatics on Mathematics and its Teaching" (Churchhouse) was also affected by a great enthusiasm concerning the perspectives of mathematics education in view of the availability of new technologies. Many mathematics educators, for instance Jim Kaput, predicted that new technologies would change all fields of mathematics education rather quickly.

> Technology in mathematics education might work as a newly active volcano – the mathematical mountain is changing before our eyes (1992, p. 515).

In the ICMI Study 17 "Mathematics Education and Technology—Rethinking the terrain" (Hoyles & Lagrange, 2010) disappointment is quite often expressed about the fact that—despite the countless ideas, classroom suggestions, lesson plans and research reports—the use of DT has not succeeded, as many had expected at the beginning of the 1990s. The disappointment that "Technology still plays a marginal role in mathematics classrooms" (ibid., p. 312) is expressed quite often. In her closing address concerning "The Future of Teaching and Learning Mathematics with DT" Michèle Artigue summarizes in the ICMI Study:

> The situation is not so brilliant and no one would claim that the expectations expressed at the time of the first study (20 years ago) have been fulfilled. (p. 464)

This study gives a good overview of the numerous activities in the last years concerning the use of digital technologies in mathematics education (also see Weigand, 2010). However, the book is not a vision; it rather poses questions, these are, however, quite similar or very similar to those 20 years before.

One may interpret that as—partial—resignation, but one can also see it as an indicator of how hard it is to answer these questions. Finally, one can also understand it as a request and as a challenge to develop new ideas—visions—in order to make progress with the integration of DT in mathematics education.

Worldwide, the current situation concerning the use of DT—and especially CAS—is very versatile. There are countries (like Norway or Denmark) that are intensively using laptops, tablets (with the programs *Geogebra* or *Maple*) or symbolic calculators (like the TI-Nspire or the Casio Classpad). These countries even allow using these tools in examinations. There are other countries (like the UK or France) that allow "only" symbolic calculators in examinations, there are countries—especially in Asia—which are very sceptical about the use in examinations, and there are countries (like Germany) where there are a different situations about the use of DT—depending on the state.

The results of the 17th ICMI-study and the ambivalent acceptance of DT in different countries made us think about the reasons for these developments (Weigand, 2013). One thesis has been that

> We underestimated the difficulties of DT-use–in a technical sense and in relation to the contents–and we have not been able to convince teachers, lecturers at university and parents of the benefit of DT in the classrooms. (ibid., p 300)

Reflecting the developments of the use of DT and especially CAS in the last decade, we started to rethink the results concerning the possibilities of supporting students' learning processes and then especially raised the question: What is the benefit of using CAS in the classroom? More specifically we asked:

1. In relation to which mathematical contexts does the CAS-use make sense and which (mathematical) competencies are supported and developed?
2. Which mathematical and tool competencies are necessary, or at least helpful, when working with CAS for specific mathematics content?
3. How can the CAS-use be described in a more detailed form?

These questions are answered based on existing theoretical considerations, on the results of the various empirical investigations and on abilities and students' competencies that are necessary to adequately work with CAS. The result is expressed in a competence model for CAS-use because an answer to these questions needs various considerations concerning the content, the tool and the way or the method of teaching and learning.

In the following we give answers to these questions, illustrated by examples, concerning three basic concepts of calculus—sequences and limits, functions and equations—and connect them to a competence model for CAS-use. These examples shall give teachers and lecturers arguments for the use of CAS in the classroom. They are not presented in the form of teaching units, but they are—of course—open for the use in calculus classes.

3 Competence Models for CAS-Use

3.1 Theoretical Foundations

The concepts of *competence* and *competence* (*level*) *models* have aroused interest in mathematics education in the past years. Starting with the NCTM Standards (1989) and especially the PISA studies, *competence* and *competencies* are expressions, often used in the context of standards and substituted the "old expression" *goals* which envisaged knowledge and abilities in mathematics education. "Mathematical competence means the ability to understand, judge, do, and use mathematics in a variety of intra- and extra-mathematical contexts and situations in which mathematics plays or could play a role …" (Niss, 2004, p. 120). In the PISA studies, competencies are on the one hand related to the *content*, e.g. numbers, space and shape, change, etc., and on the other hand—in a more general way—related to *processes* like problem-solving, modelling and the use of mathematical language. In order to evaluate or operationalize the competencies through the construction of items and tests, it is helpful to organize these competencies in levels, categories or classes. In the PISA studies, each of the possible pairs (content, process) can be divided into three different levels or competence classes (OECD, 1999, p. 43):

- Class 1: reproduction, definitions, and computations
- Class 2: connections and integration of problem solving
- Class 3: mathematical thinking, generalisation and insight.

This leads to a three-dimensional competence-model with the *dimensions* content, basic or process competencies and cognitive activation.[5] The German Standards for grade 10 (KMK, 2004) and grade 12 (KMK, 2012)—based on the NCTM Standards—are also related to *content*[6] and *general competencies.*[7] And there are also three levels of cognitive activation, which are traditionally called *Reproduction*, *Connections* and *Reflections*, these levels are in line with the PISA classification.

3.2 Competence Model for CAS-Use While Working with Functions

In Weigand and Bichler (2010a) a Competence Model for the Use of Symbolic Calculators in Mathematics Lessons in the frame of working with functions was developed. Different levels of understanding the function concept have been seen in relation with the "tool *competencies*" *and—as a third dimension—with the* "*cognitive activation*" (Fig. 1).

The ability or the competence to adequately *use the tool—here a CAS*—requires technical knowledge about the handling of the tool. Moreover, it requires the knowledge of when to use which features and for which problems it might be helpful. The *use of SC* was classified according to the way representations are used. We distinguished three levels, which might also be categorized by using SCs—Sientific Calculators or Graphing Calculators with CAS—as a (simple) function

[5]These dimensions are in PISA called "Overarching ideas" (content), "Competencies" (process) and "Competence Clusters" (cognitive activation).

[6]These are: numbers, measuring, space and shape, functional connections, data and chance.

[7]These are: arguing mathematically, solving problems mathematically, modelling mathematically, using mathematical representations, acting mathematically on a symbolical, formal and technical level, communicating mathematically.

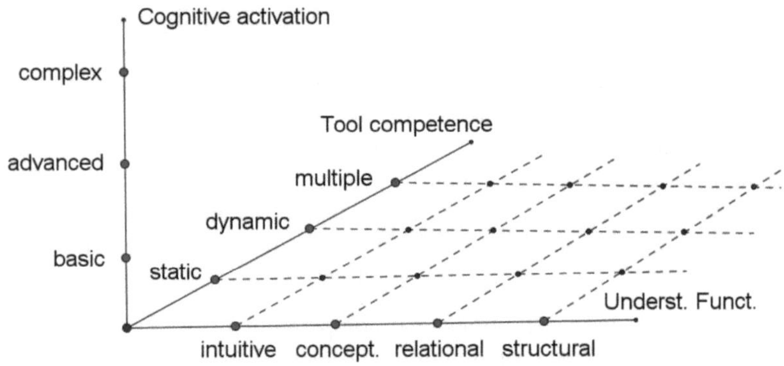

Fig. 1 Competence model for CAS-use while working with functions

plotter, as a tool for creating dynamic animations and as a multi-representational tool.

The competence model had a diagnostic goal to describe the anticipated aim in the context of the project, but also to detect, which abilities are missing while students are working with SCs. However, the evaluation or the diagnosis can only be the very first step. The second step is to think about consequences based on the evaluation. How can the student be supported to reach higher levels of understanding and higher competence levels while working with the SC?

3.3 Competence Model for CAS-Use

In the following, the competence model in 3.2 is extended and specified on the content dimension in the frame of calculus. In school mathematics, calculus is the field, where CAS are mostly used, compared to geometry or stochastics. The first dimension "Understanding functions" was substituted by three basic concepts of calculus: sequences and limits, functions and equations. The second dimension "Tool competence" was substituted by "Representation" with the categories "isolated", "multiple static" and "multiple dynamic" because this matches real working with CAS better and is on the same level as the theory of representation, which emphasizes the reasoning with multiple and dynamic representations (see Bauer, 2013 or Ainsworth, 1999). Moreover, the concept "Tool competence" rather describes a competence, which should be seen in relation to working with special contents and representations or even also including classroom activities. Thus, the expression "Tool competence" is not used as a dimension in the competence model, but it is rather a competence, which includes the here discussed competencies. The third dimension "Cognitive Activation" was substituted by "Activity" and emphasizes the classroom activities in which the CAS is used more:

- Calculate: the CAS as a tool for (numeric and symbolic) calculations;
- Consult: the CAS as a formulary;
- Control: the CAS as a controller of hand-written solutions, suggestions and ideas;
- Communicate: the CAS as a source for explanations and argumentations—especially if it is used as a "black box";
- Discover: the CAS as a tool for evaluating and testing suggestions and strategies in a problem solving process.

This classification may be seen as a hierarchy while moving from a procedural knowledge (calculate) to a conceptual knowledge (communicate, discover). This new third dimension is seen more on the teaching side while the dimension "Cognitive activity" (in model 3.2) is more on the learning side. It would be possible to add the "Cognitive activity" as a fourth dimension, but that would increase the complexity of the model. We left this fourth dimension out, because the model is already quite complex and seems, in a first approach, to concentrate on problems in relation to the three dimensions and later to think about different levels of difficulties with these problems.

The present three-dimensional competence model—represented in a three-dimensional coordinate-system—has three categories on the topic- or content-axis, three categories on the representation-axis and five on the activity-axis. This gives us $3 \times 3 \times 5 = 45$ cells. If each cell is again subdivided into three levels of cognitive activation, this makes a total of 135 cells (Fig. 2).

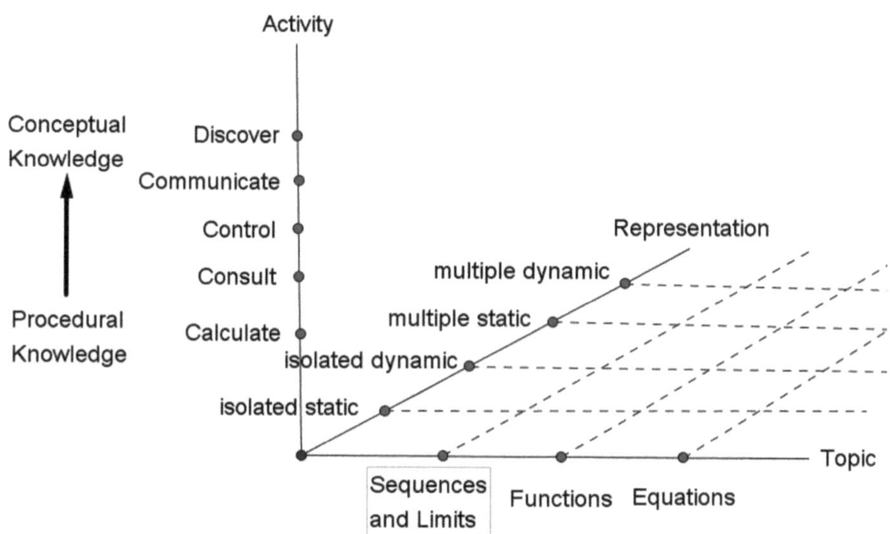

Fig. 2 An extended competence model for CAS-use while working with sequences, functions and equations

When we speak of a CAS, we think of a tool which is able to represent mathematics or mathematical objects on symbolic, graphic and numeric level. We use the word *tool* instead of *instrument* because we concentrate on the facilities of CAS, reflect these in relation to mathematics aspects in the classroom. However, we leave the development of the user-tool-relationship in the frame of an instrumental orchestration, which is the heart of the instrumental genesis (see Artigue, 2002; Drijvers et al., 2010), to the user or learner.[8]

4 Examples

In the following, we will concentrate on the relation of "topic" and "representation" and present examples that are typical of the particular "cells". It will not and cannot be possible to assign particular examples to one type of representation. However, there will be fluent transition between isolated und multiple as well static and dynamic representations. The focus will of course be on the significance of the CAS representations for the different topics.

4.1 Sequences

4.1.1 Recursively Defined Sequences

Sequences whose elements can be defined by the previous elements in a straightforward way are called *recursively-defined* sequences. With a first element $a_1 \in \mathbb{R}$ and a function f: $\mathbb{R} \to \mathbb{R}$ is then $a_{k+1} = f(a_k)$ with $k \in \mathbb{N}$ and we get a *sequence of iteration*

$$a_1, a_2 = f(a_1), a_3 = f(a_2), \ldots$$

Figure 3 shows the numerical calculation of the first few elements of

$$f(x) = -0.6 \cdot x + 3 \text{ with } a_1 = 1$$

in the isolated representation of a spreadsheet (Fig. 3). In the following we mainly use the program Geogebra, which is a combination of a CAS, function plotter, dynamic geometry software and a spreadsheet, and the Casio ClassPad either as a symbolic calculator or a notebook simulation.

In the CAS window, these numerical calculations can be represented on a symbolic level in a way that is closer—but not close—to the traditional mathematical notation. However, one has to be familiar with the particular commands in the "language of tools" (Fig. 4).

[8]In German language the concepts *tool* and *instrument* are widely used interchangeably.

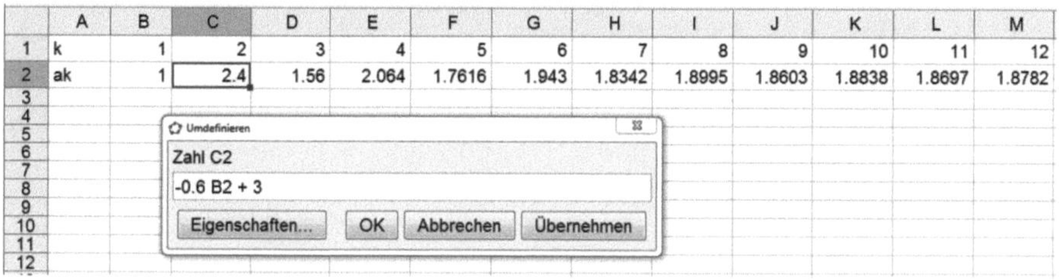

	A	B	C	D	E	F	G	H	I	J	K	L	M
1	k	1	2	3	4	5	6	7	8	9	10	11	12
2	ak	1	2.4	1.56	2.064	1.7616	1.943	1.8342	1.8995	1.8603	1.8838	1.8697	1.8782

Fig. 3 Table view (Geogebra) of the function $f(x) = -0.6 \cdot x + 3, a_1 = 1$

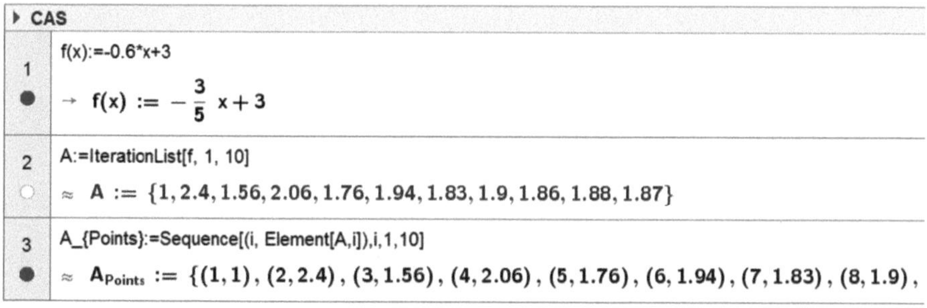

Fig. 4 Calculation (Geogebra) of the sequence elements of $a_{k+1} = -0.6 \cdot a_k + 3, a_1 = 1$ in the CAS window

Changing the original function f leads to an automatic change of the numerical values.

Recursively defined functions can be well represented in a "cobweb-diagram". The following representation (Fig. 5) with a given function $f{:}x \to y, x \in \mathbb{R}$ and the first element a_1 shows the schematic construction of the iteration sequence.

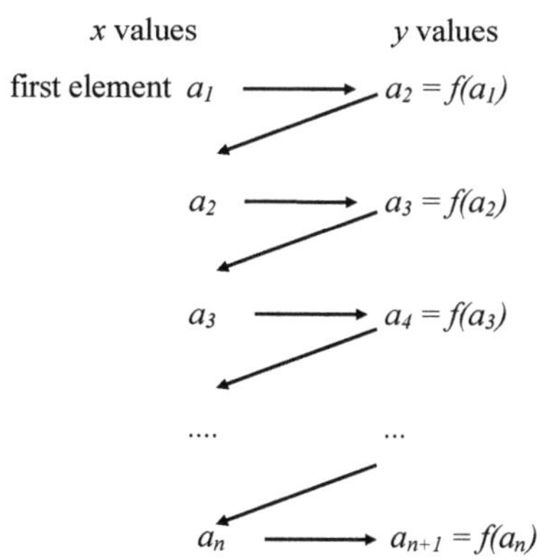

The convergence of the function for "large k" can be seen in a "contraction" of the sequence points in the intersection of the line $f(x) = -0.6 \cdot x + 3$ with the angle bisector of the first quadrant with $y = x$.

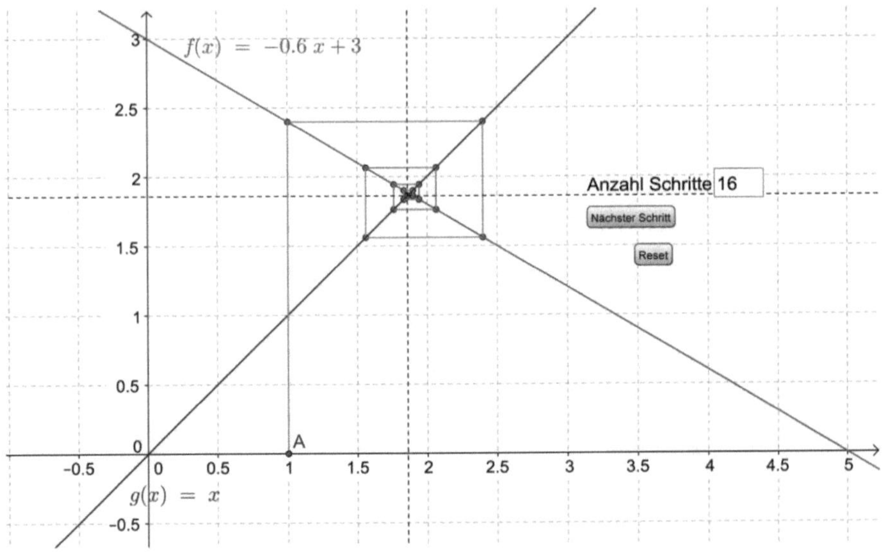

Fig. 5 The “cobweb-diagram” (Geogebra) of the sequence $a_{k+1} = -0.6 \cdot a_k + 3, a_1 = 1$

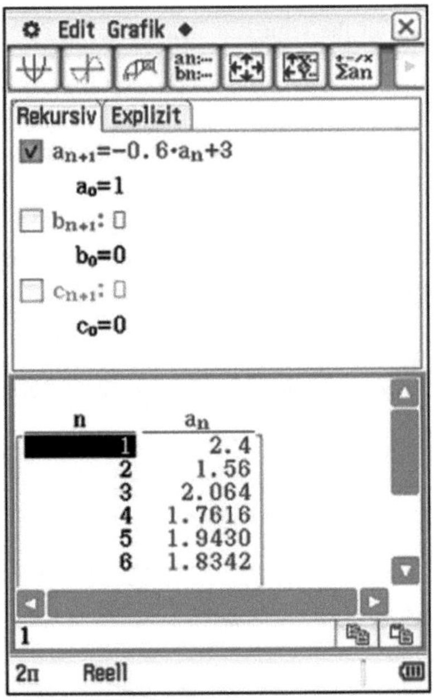

Fig. 6 Symbolic and table representation (ClassPad) of a recursively-defined sequence

With the Casio ClassPad the notation of the iteration sequence is quite close to the mathematical language (Figs. 6 and 7).

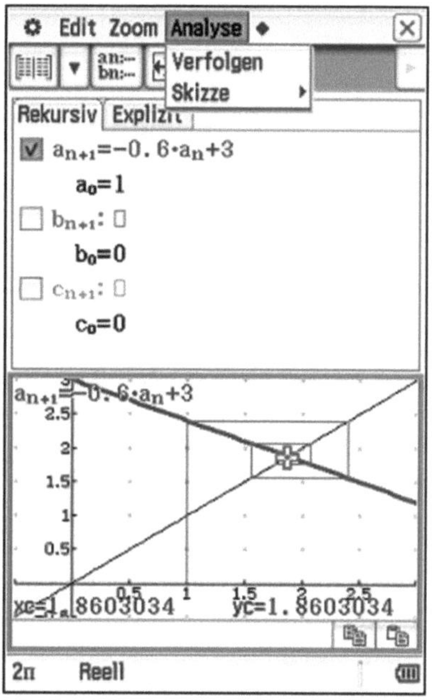

Fig. 7 The cobweb-diagram (ClassPad) of a recursively-defined sequence

In this case, CAS is a tool that helps showing the sequence on a symbolic level, enables numerical calculations, provides the (numerical) basis and is a device for experimentation in the field of discovering relations. The first example, as well as the following ones, shows that CAS has to be seen and evaluated in relation with other representations, especially in the relation with a heuristic examination and discovery. Furthermore, the examples underline the importance of the dynamic aspect.

Over all some or many or nearly all aspects of the mathematical concept of a sequence are seen in relation to the representations and some activities of the competence model for CAS-use (see Sect. 3.3). The CAS does calculations, supports communication (between mathematics and the user, but it is also a medium concerning the teacher-learner- and the learner-learner-communication), and it is a helpful tool in the discovery process (Fig. 8). Especially the following mathematics or calculus competencies—beyond the CAS-use—are supported: Learners

- recognize the unrestrained continuation of an "infinite processes", which is the possibility of coming arbitrarily close to a "limit object";
- combine graphical and numerical ideas with the limit processes in the sense of "pursue to" or "is arbitrarily close to";
- can numerically and graphically describe the limit process with recursively defined sequences with $a_{k+1} = A \cdot a_k + B, A, B \in \mathbb{R}, A = const, B = const.$

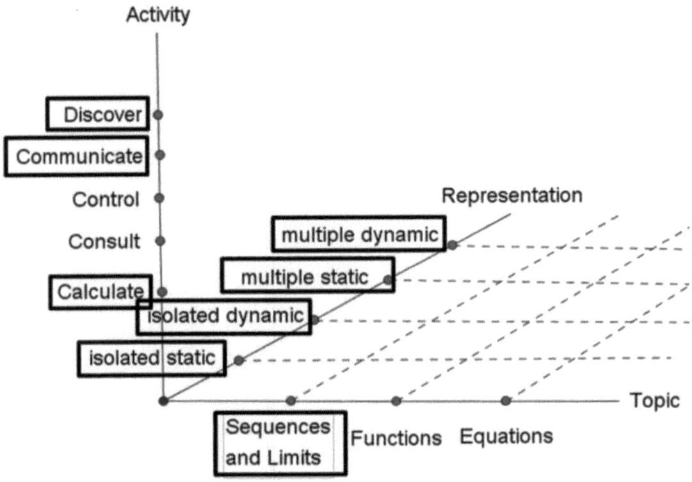

Fig. 8 Competence model for CAS-use while working with recursively defined sequences

4.1.2 Difference Sequences

Difference sequences $(\Delta a_k)_{\mathbb{N}}$ with

$$\Delta a_k := a_{k+1} - a_k$$

with a given sequence $(a_k)_{\mathbb{N}}$ are well suitable for a discrete introduction of the difference quotient. Based on sequences or functions that are defined on $\mathbb{N}$, we will now take a look on functions $f:\mathbf{Z} \to \mathbb{R}$, the so called *Z-functions* that are defined on Z and their relation with difference-Z-functions. $D_f : D_f(z) = f(z+1) - f(z)$, p.e. $f(z) = z^2 - 2z + 3$ (Figs. 9 and 10).

The dependence of D_f on the used parameters of f with $f(z) = az^2 + bz + c$ can be graphically depicted. The dynamics of the representation can be induced by the "slide bars" (Figs. 11 and 12).

The two graphs of f and D_f already suggest that the graph of the difference-Z-function is linear. This can be explained on a symbolic level: With $f(z) = az^2 + bz + c$ you receive the difference-Z-function

$$\begin{aligned} D_f(z) &= f(z+1) - f(z) = a(z+1)^2 + b(z+1) + c - \left(az^2 + bz + c\right) \\ &= 2az + a + b. \end{aligned}$$

Therewith, the changes of the graph by varying a and b and the independence of D_f can be explained. Moreover, it has to be noticed, that the mathematical communication—concerning used expressions and gestures—is different in static and dynamic environments. You will get more dynamic verbalizations if you use dynamic representations (Ng, 2016).

Fig. 9 $f(z) = z^2 - 2z + 3$

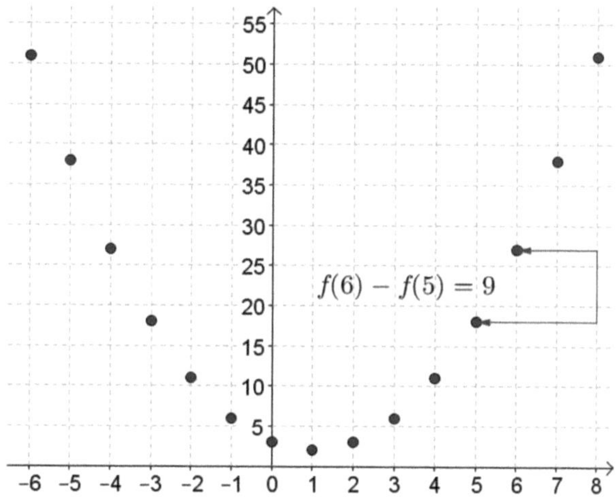

Fig. 10 $D_f(z) = f(z+1) - f(z)$

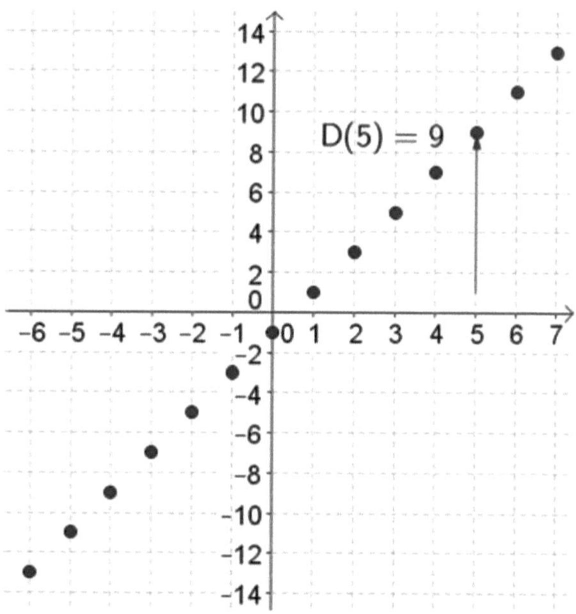

This concept of the Z-function can be transferred to polynomials of a higher degree. For example, for the Z-polynomial with

$$f(z) = az^3 + bz^2 + cz + d$$

you receive the difference-Z-polynomial

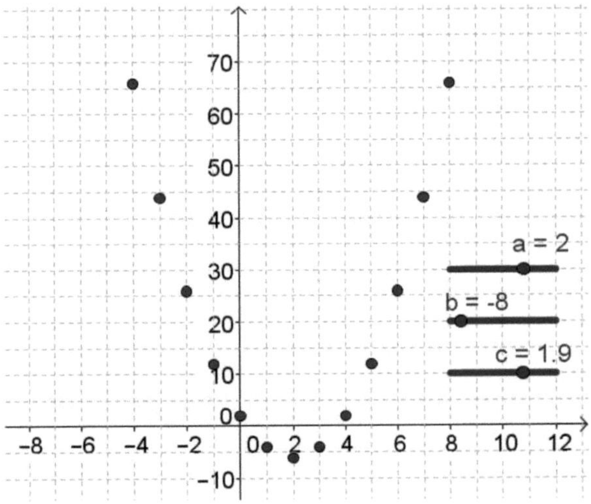

Fig. 11 $f(z) = z^2 - 8z + 1.9$

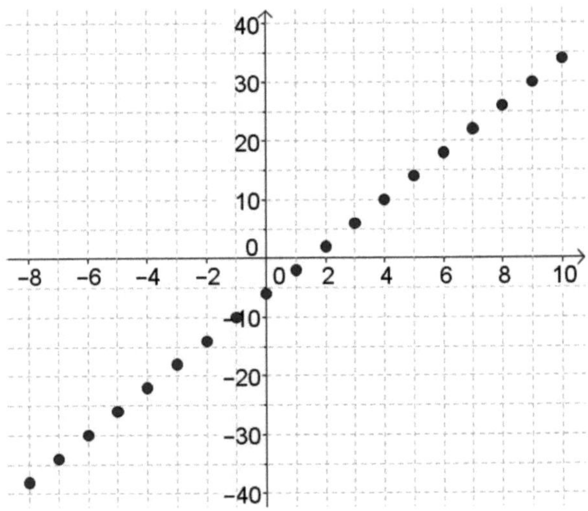

Fig. 12 $D_f(z) = f(z+1) - f(z)$

$$D_f(z) = 3az^2 + (3a + 2b)z + a + b + c.$$

This can easily be calculated with a CAS or at least be verified with the programme (Fig. 13).

The use of computer algebra systems is especially useful and helpful when difference-Z-polynomials of Z-polynomials of a higher degree have to be calculated (Fig. 14).

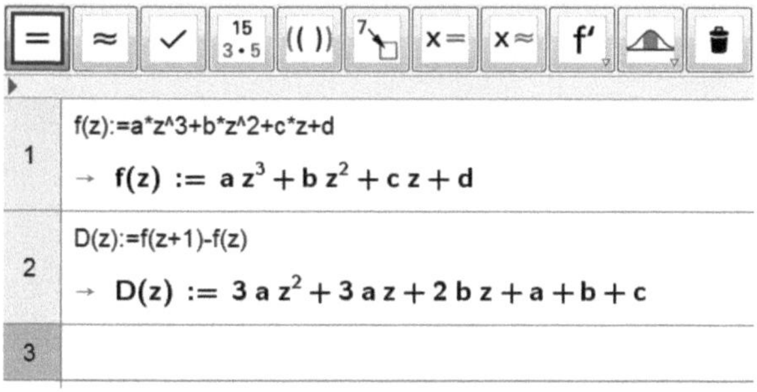

Fig. 13 Calculation (Geogebra) of the difference-Z-function with a CAS

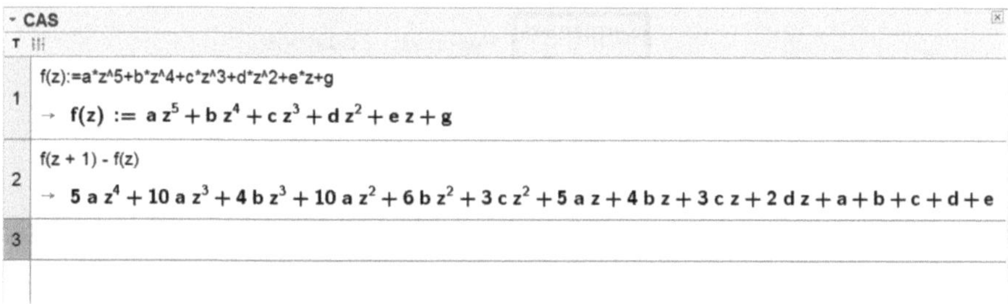

Fig. 14 Calculation with a CAS

This leads to the hypothesis that a Z-polynomial of degree n ($\in \mathbb{N}$) has a difference-Z-polynomial of degree n-1.[9]

Here we can see that there are different aspects or representations that are connected to the CAS. It is a tool for multiple and dynamic representations, it is an experimentation tool and reflects mathematical expressions on a symbolic level with a notation that is close to mathematical notation. Concerning the competence model for CAS-use, all kinds of representations are used and also nearly all kinds of activities—except consult—are already used in these examples (Fig. 15).

We summarize the meaning of CAS concerning the content *sequences and limits*:

- CAS is a tool with notations (or a language) quite close to mathematical notations (or the mathematical language);
- CAS allows object-related working with sequences and discrete functions;
- CAS has to be seen or evaluated in relation to other—especially graphical—representations.

[9]For G.W. Leibniz (1646–1715) sequences and their difference sequences have been a source for the development of the derivative and the calculus.

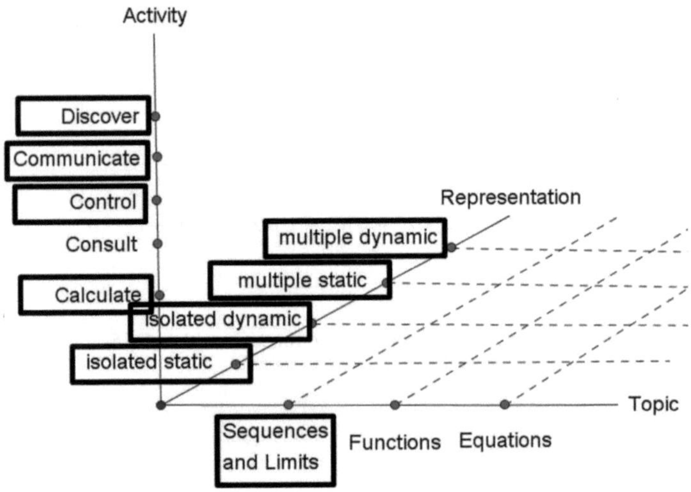

Fig. 15 Competence model for CAS-use while working with difference sequences

Especially the following mathematics or calculus competencies are supported: Learners

- understand the definition of difference sequences of sequences and Z-functions and realize the relation between a Z-function and its difference sequence;
- interpret the relation between a Z-function and its difference sequence in different representations;
- can determine the difference-Z-function of a Z-function on the symbolic level, basic examples by hand and more complex examples using a CAS.

4.2 Functions

4.2.1 Working with Functions as Objects

Functional thinking expresses itself in a way of thinking that emphasizes three aspects of the function concept (cf. Vollrath, 1989):

- The aspect of allocation: A function as an allocation of (particular) values.
- The aspect of change: A change of the independent values leads to a change of the dependent values.
- The aspect of object: A function is an object that depicts a connection as a whole.

This last aspect is especially important, when functions are being added, subtracted, multiplied, divided or composed. A special case is the change of the function.

$f{:}x \to y = f(x)$ with $a, b, c, d \in \mathbb{R}$ to $a \cdot f(x), f(x) + c, f(x + d), f(bx)$ or $f(bx + d)$

Therewith, a change $f(x) \to a \cdot f(x)$ can geometrically be described as "enlarging" (or shrinking) in y-direction with the function

$$A : \mathbb{R}^2 \to \mathbb{R}^2, \begin{pmatrix} x \\ y \end{pmatrix} \mapsto \begin{pmatrix} 1 & 0 \\ 0 & a \end{pmatrix} \cdot \begin{pmatrix} x \\ y \end{pmatrix} = \begin{pmatrix} x \\ ay \end{pmatrix}.$$

The change $f(x) \to f(bx)$ can be described as enlarging (or shrinking) in x-direction by the factor $\frac{1}{b}$ with

$$B : \mathbb{R}^2 \to \mathbb{R}^2, \begin{pmatrix} x \\ y \end{pmatrix} \mapsto \begin{pmatrix} \frac{1}{b} & 0 \\ 0 & 1 \end{pmatrix} \cdot \begin{pmatrix} x \\ y \end{pmatrix} = \begin{pmatrix} \frac{1}{b}x \\ y \end{pmatrix}.$$

Figure 16 shows these maps of the graph G_f with $f(x) = \frac{1}{3}x^3 - 2x$.

In the CAS window, the corresponding transformations with the parameters distance can be made on a symbolic level. Hereby, one has to consider if a parameter, e.g. a, is looked at "in general", or if particular values are being calculated e.g. for drawing the graph. Therefore, it is useful to take a look at functions that depend on parameters as functions of several parameters and to indicate this in the definition of the function, e.g. $g(x, a){:=}a \cdot f(x)$. The graph of the function g can then be drawn for a particular value a (here named "aa") with the function $G_g(x) = g(x, aa)$. The CAS requires on the one hand, and probably visualizes on the other hand, where the learner has to distinguish mentally a parameter as a general variable, or as variable with a particular value, when working traditionally with paper and pencil.

4.2.2 Milk Packs

For the following considerations concerning the optimization of the material consumption of a milk pack, the package of Fig.17 (c) will be used. At first, it is surprising that the net is a square having sides of 28.5 cm. The question arises, if this pack has only been constructed for optimality reasons, i.e. if the packaging material for the volume of 1 L is minimal regarding the surface.

For the actual milk pack, one has h = 18.8 cm, b = 7.1 cm. If you open a milk pack that is still closed at its cap, you will notice that the milk fills the pack more or less up to the height h, therefore, there is still a bit of air in the "roof". The volume can be calculated with the help of the parameters that are listed in Fig. 18, as follows:

$$V(b, h) = b^2 \cdot h = 1000\left[\mathrm{cm}^3\right]$$

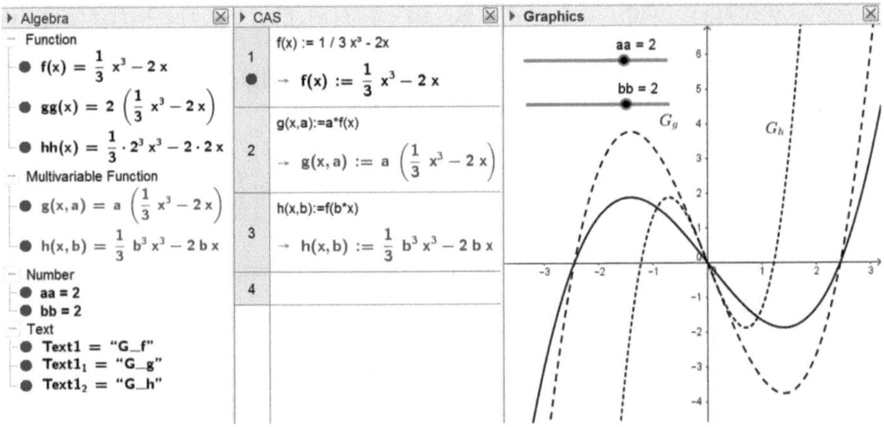

Fig. 16 The graphs (Geogebra) G_f, G_g, G_h with $f(x) = \frac{1}{3}x^3 - 2x$ and $g(x) = a \cdot f(x)$ and $h(x) = f(b \cdot x)$

and

$$h = \frac{1000}{b^2}.$$

For the lateral area results:

$$M(b,h) = (4b + 0,5)(h + 1.7 + b + 2).$$

The CAS does not only perform the particular calculations, but also provides a good and clear structure of the solution of the task (Fig. 19).

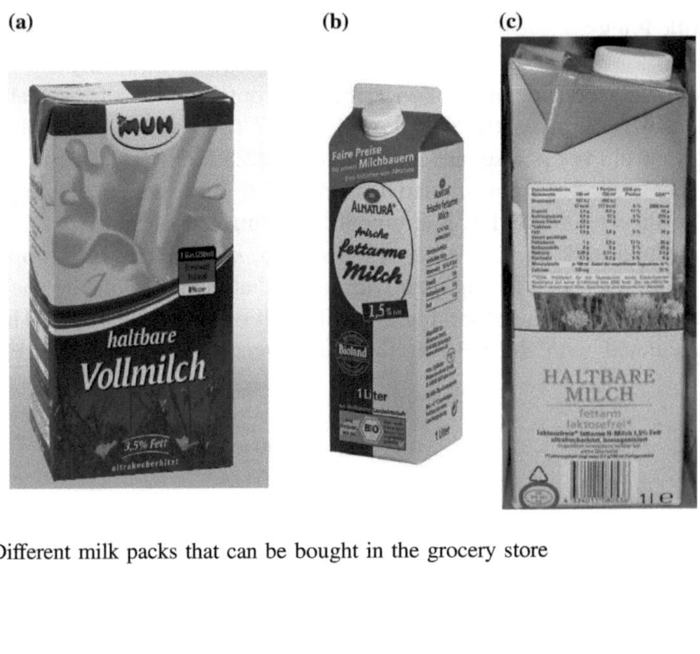

Fig. 17 Different milk packs that can be bought in the grocery store

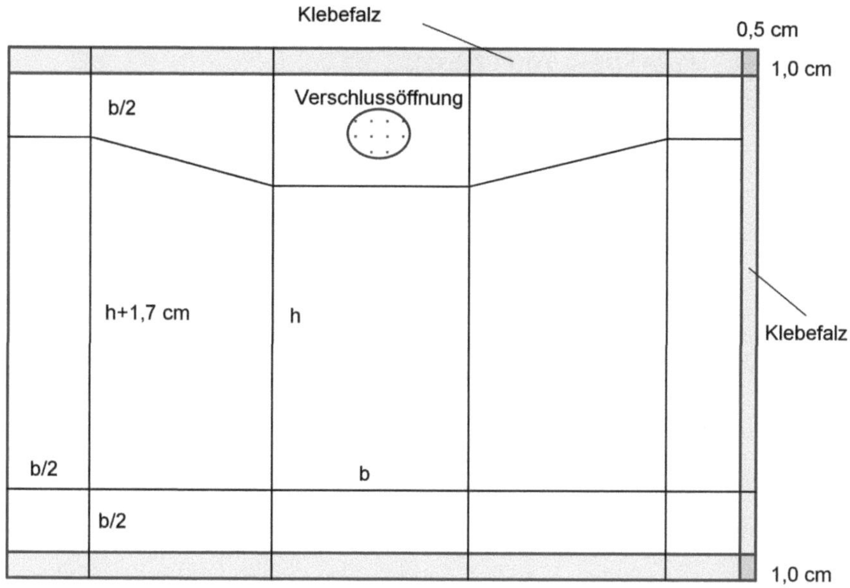

Fig. 18 Net of the milk pack from Fig. 17(c)

The CAS is in this case at first solely being used statically for the actual case. For an optimal width of the pack, one has a value of $b = 7.44$, when two decimal places have been adjusted. This can also be represented graphically and the optimal b-value can be approximately determined from the graphic (Fig. 20).

When handling the results that had been obtained through the mathematization experimentally, the width of the fold f—measured with 1 cm at the original pack—can be changed and the effects on the optimal b-value can be studied. With the help of a slide bar, the f-values can be changed easily. Therewith, one has an isolated dynamic symbolic representation (Fig. 21).

Another possibility would be to leave the value f as a general parameter in the original equation (Fig. 22).

Here, Geogebra reaches its limits, as Eq. 4 cannot generally be solved anymore with f as *a* parameter.

Additional remark to the modeling problem: With the model above, one receives an optimal value for $b = 7.44$ cm regarding the surface of the pack. It can now be discussed why this value differs—also if only a bit—from the actual value. Production conditions (e.g. width of the cardboard rolls, the packs are cut out of) can be named, or it is remarked that the sides of the pack are not even, when the pack is filled with milk!

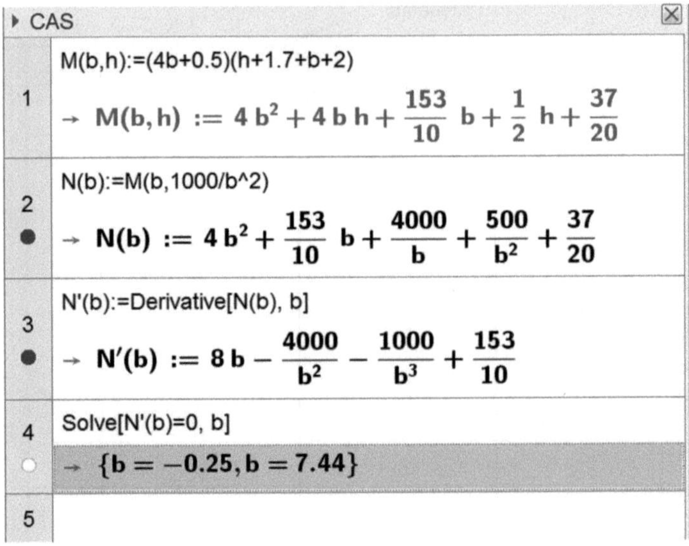

Fig. 19 Calculation (Geogebra) of the length b of the milk pack

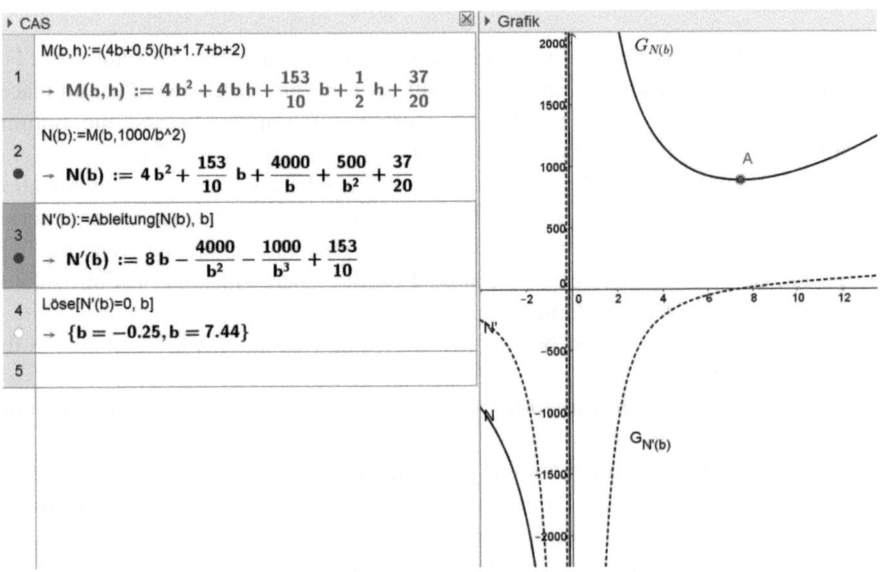

Fig. 20 Symbolic and graphic representation (Geogebra) of the optimization of the milk pack

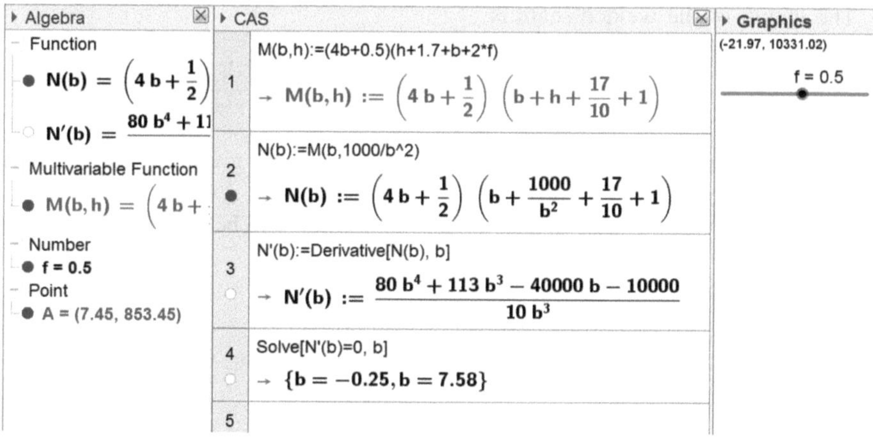

Fig. 21 Calculation (Geogebra) of the optimal pack with a variable width of the fold f

4.2.3 Welded Seam of a Tin Can

The following considerations ask for the optimization of a welded seam of a can with volume V (Figs. 23 and 24). There is a welded seam along the circular bases on bottom and top and along the height h. If we use a cylinder with radius r and height h as a mathematical model, the volume V of the can is:

$$V(r, h) = r^2 \cdot \pi \cdot h.$$

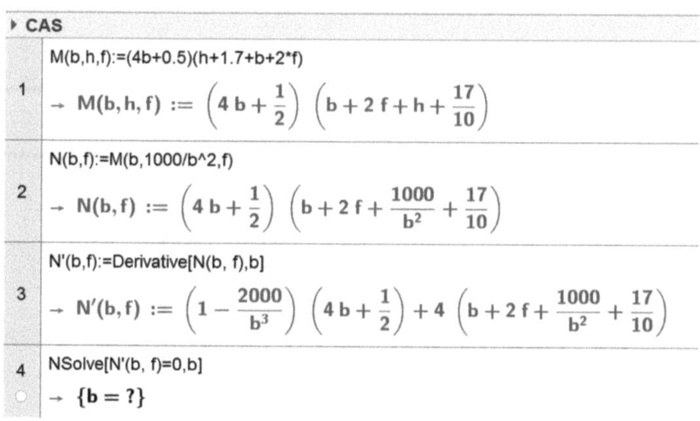

Fig. 22 General calculation with the width of the fold f

The length of the welded seam is:

$$L(r, V) = 2 \cdot (2 \cdot r \cdot \pi) + h = 4 \cdot r \cdot \pi + \frac{V}{r^2 \pi}.$$

Is there a minimal length of the welded seam with the given volume V. What does the can look like? (Fig. 25).

If we use the Casio ClassPad also the last expression will be simplified (Fig. 26):

Fig. 23 A tin can

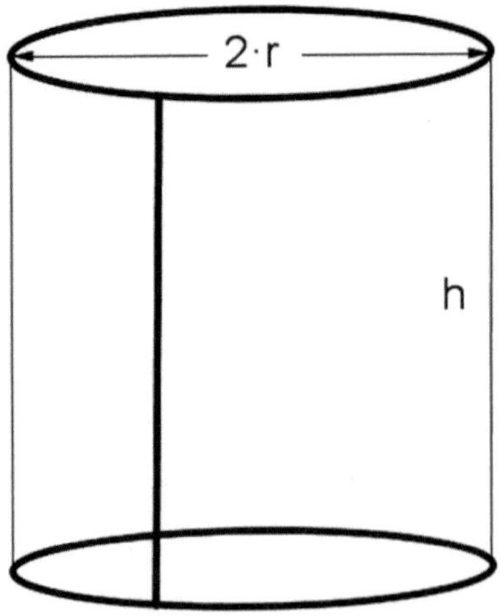

Fig. 24 The welded seam of a tin can

The ratio of the height and the diameter of the optimized can (concerning the length of the welded seam) is π. The CAS is a tool for calculations with a static symbolic representation and it is a multiple and dynamic tool for representing special examples—with a "slider"—on the graphic level.

While working with functions and using the CAS we work on different levels or in different "cells" of the competence model. Calculations are mainly done isolated in the symbolic representation. For discovering, multiple dynamic representations are used in the graphic representation in relation to the symbolic representation. Moreover, a CAS can be used as a control tool in the discovery process.

We summarize the meaning of CAS concerning the content *functions:*

- CAS allow calculations on the symbolic level in notations which are quite close to the common mathematical notation or language.
- CAS expand the working with functions as objects on the symbolic level and give the possibility of a simultaneous connection to numeric and graphic representations;

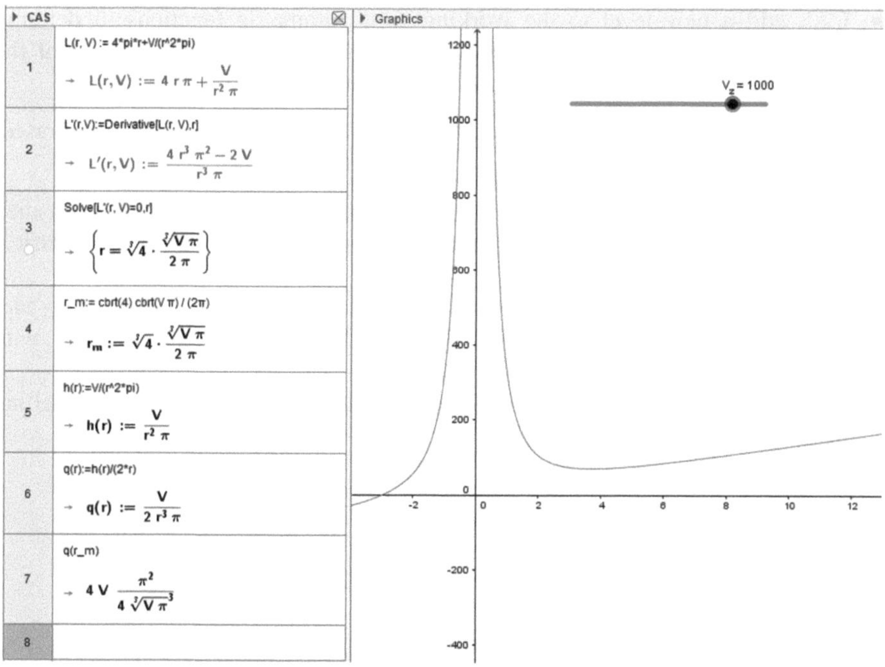

Fig. 25 Calculation (Geogebra) of the minimal welded seam length in dependence of the volume V

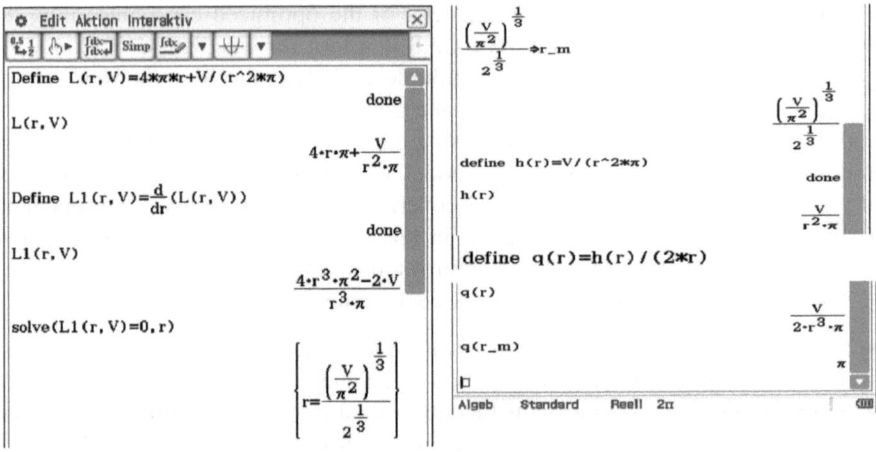

Fig. 26 Calculation (ClassPad) of the minimal welded seam length in dependence of the volume V

- CAS add a new level to the working with parametric functions in defining functions of two (several) variables and allowing the dynamic change of the variables in the graphic and numeric representations;

Especially the following mathematics or calculus competencies are supported: Learners

- can work with functions as objects on a symbolic and a graphical level; they especially interpret changes of the variables of a function a geometrical transformations;
- understand the definition of functions of several variables and they can—adequate to the situation—interpret them as functions of one variable with parameters;
- can use functions of several variables to solve mathematical and modelling problems.

4.3 Equations

4.3.1 Linear and Quadratic Equations

A CAS is a formulary that offers in particular solution formulas for linear and quadratic equations and for systems of linear equations (Fig. 27).

In relation with a graphic representation, questions concerning the *number* of zeros of a quadratic function can (at first) be answered through experimental exploration (Fig. 28).

A CAS can be used to calculate the zeros of a function by only pressing one button, but moreover, it serves as visualization. Furthermore, the relation of function and equation is fundamental for the mutual representation in the CAS and the graphic window.

4.3.2 Systems of Equations

The CAS is a tool for solving systems of equations with parameters (Fig. 29). Also systems with quadratic equations can be calculated (on a symbolic level) (Fig. 30). The CAS provides the calculations and the solutions on the symbolic level and these have then to be interpreted, especially in relation to the graphical level.

4.4 Complex Equations

Equations of higher degree, particularly equations of degree 3, can be solved in a symbolic form by a CAS, but the solution depends on the equation (Fig. 31).

However, there will be different forms of solutions if the CAS solves the following equation (Fig. 32). The Casio ClassPad, for example, can also solve equations that depend on parameters (Figs. 33 and 34). However, a constructive handling of these solutions asks for further knowledge concerning the solution

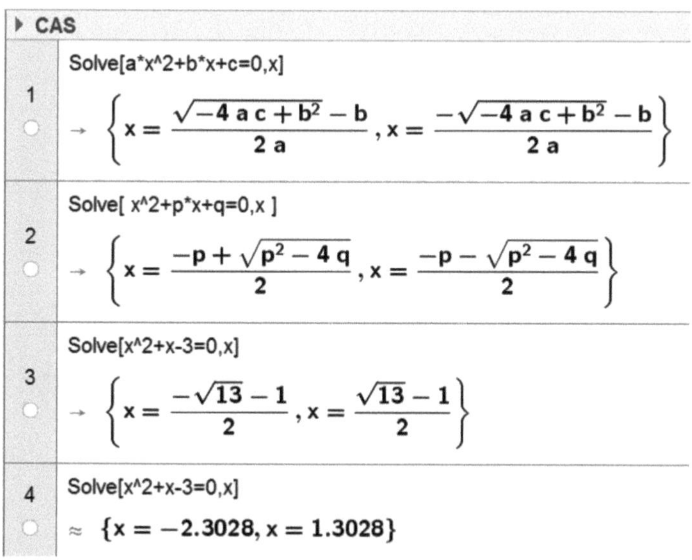

Fig. 27 Quadratic equations with CAS (Geogebra)

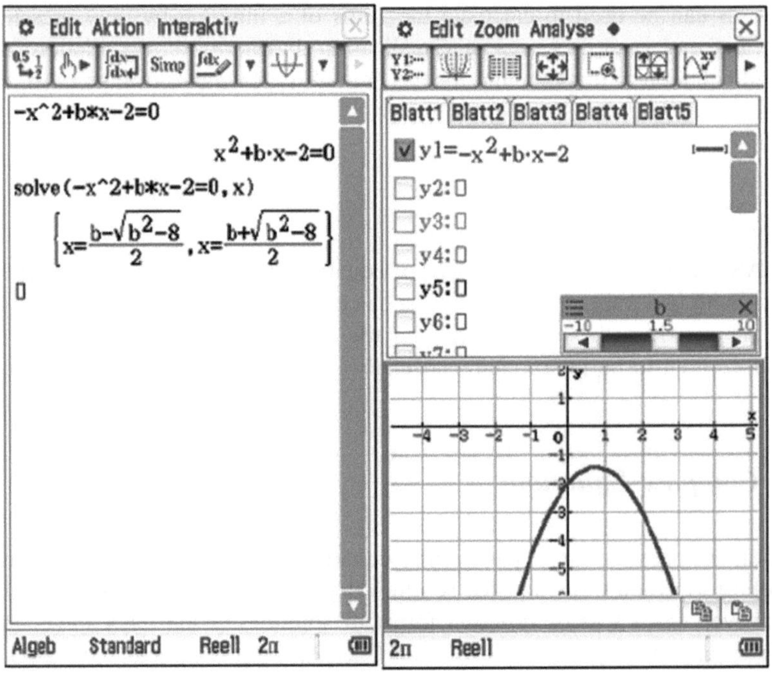

Fig. 28 Symbolic and graphic representation (ClassPad) of the solutions of a quadratic equation with a parameter

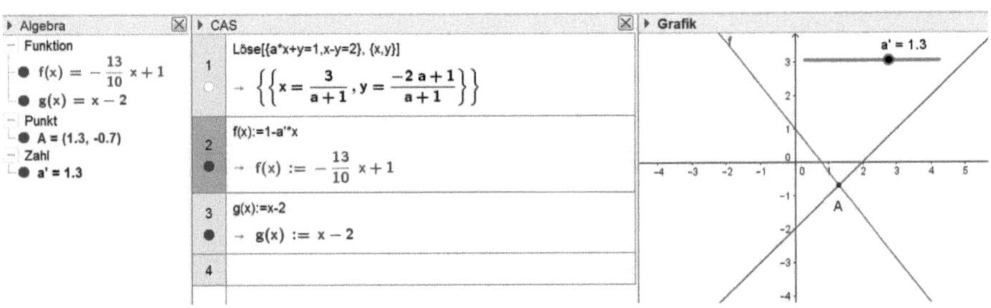

Fig. 29 Solving a linear system of equations with one parameter (Geogebra)

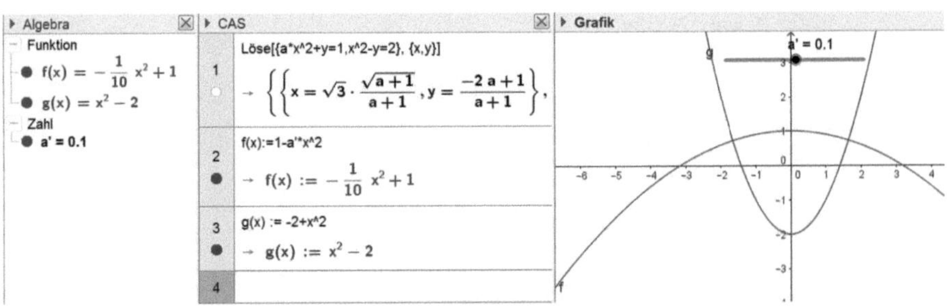

Fig. 30 Solving a system of quadratic equations with one parameter

formulas or these solutions can encourage dealing with solution formulas—here the Cardano formulas.

It is, however, only possible to solve more complex equations with a CAS, when there is already a basic knowledge of the solution variety of the considered equations. Furthermore, one needs strategies for the handling of a representation type especially, with regard to necessary changes of the representation types. Because, if an approach that had been used did not lead to a successful solution a strategy is needed. An example is the solution of the equation $1 + \sin(x) = 2^x$ (Fig. 35).

Geogebra-CAS cannot solve the equation on a symbolic level. The Casio ClassPad offers several numerical solutions (Fig. 28), these are however not easy to understand for (almost) every user (Fig. 36).

A useful strategy would here be the change to a graphic representation and zooming in on the intersection point of the graphs. Therefore, mathematical knowledge about basic properties of the two functions is absolutely necessary. Tonisson (2015) gives a good overview of the solution variety of equations, as he has solved and compared 120 equations of school mathematics with 8 different CAS.

A last example: $x^7 - 4x^5 + 4x^3 = 0$.

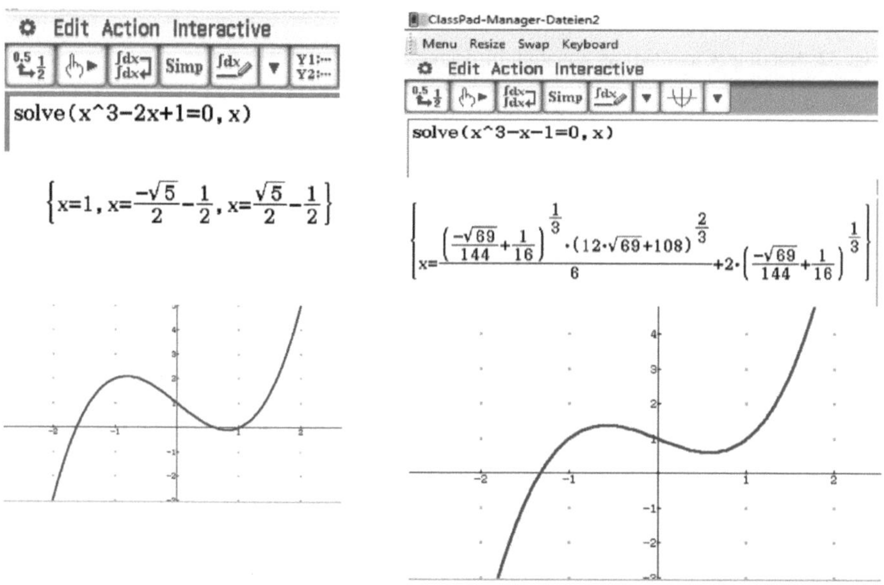

Fig. 31 Soving polynomial equations (ClassPad) of degree 3

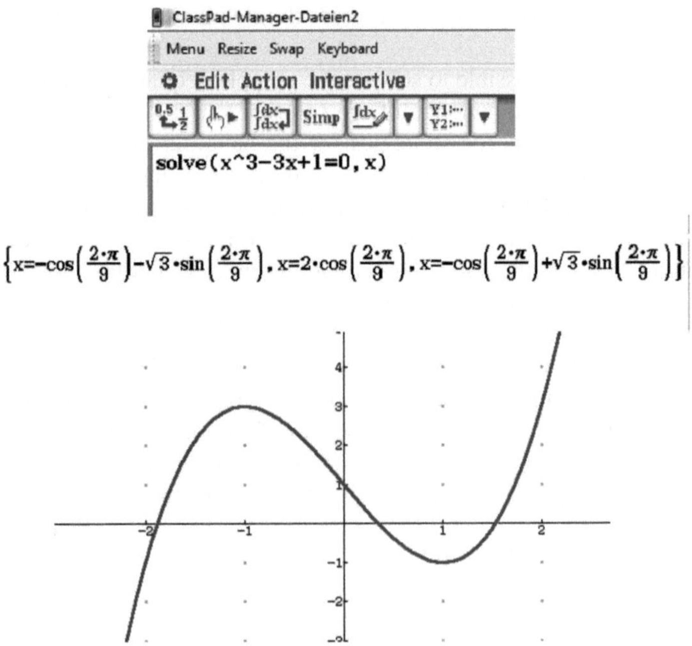

Fig. 32 Solving (ClassPad) an equation of degree 3

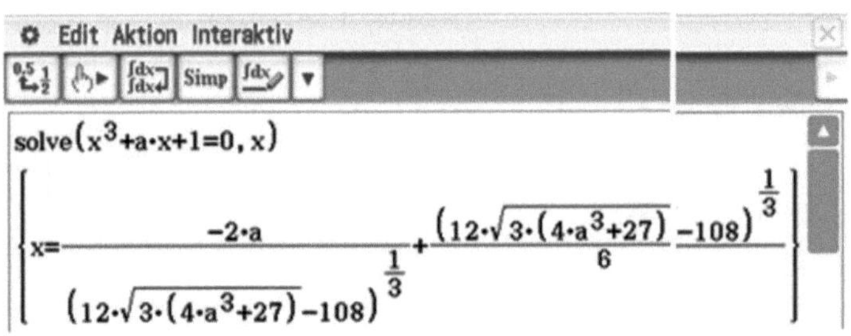

Fig. 33 Section of the solution of the equation $x^3 + a \cdot x + 1 = 0$ with the Casio ClassPad

Fig. 34 Section of the solution of the equation $x^3 + p \cdot x + q = 0$ with the Casio ClassPad

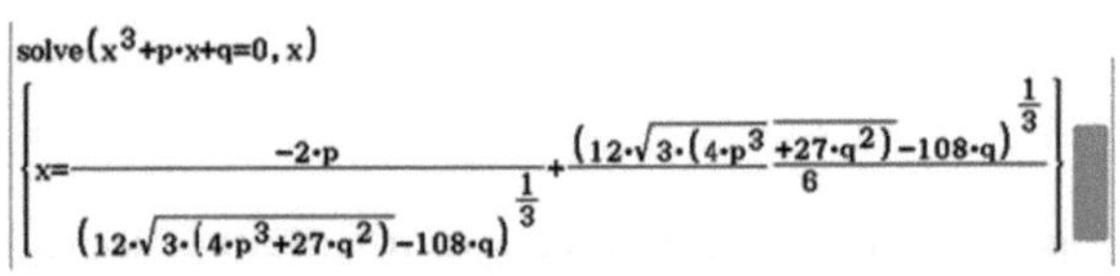

The CAS gives the solutions of a polynomial of grade 7, but only because the expression can be factorized (Fig. 37). The—surprising—solution has to be interpreted with the graphic representation.

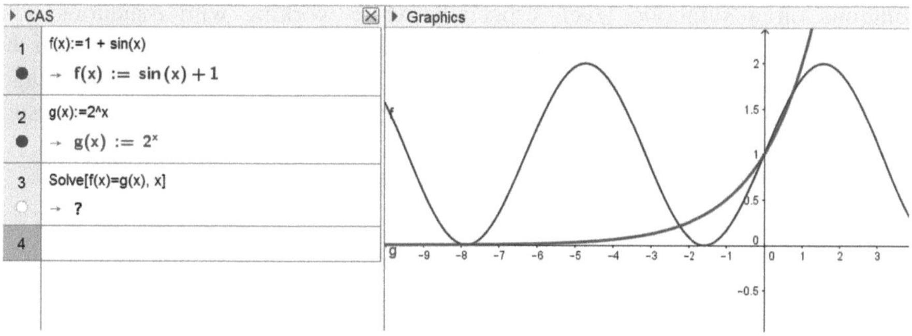

Fig. 35 Graphical solution of "complex" equations

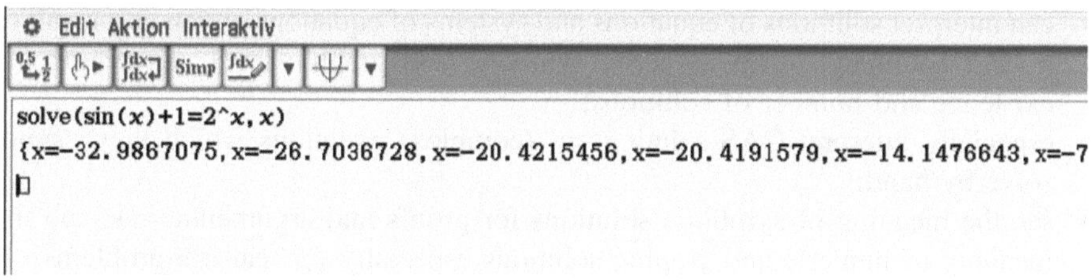

Fig. 36 Symbolic solution with the Casio ClassPad

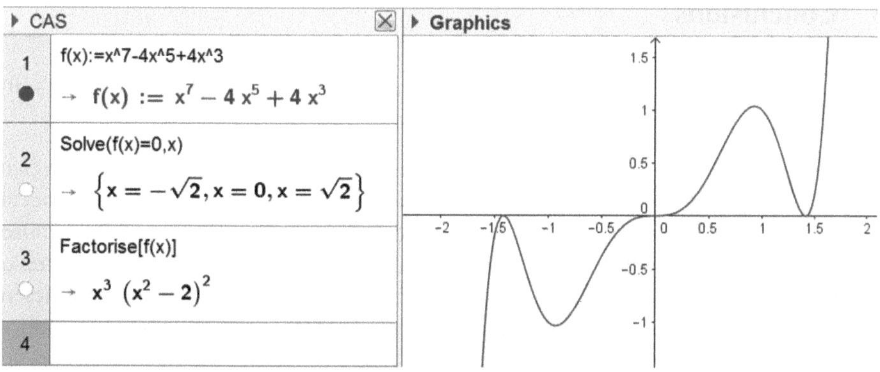

Fig. 37 Graphic solution of a polynomial of grade 7

An efficient use of a CAS when solving—a bit more complex—equations is only possible with a mathematical knowledge concerning the solution of equations, the characteristics of the underlying functions of the equations and the possibilities of the solution varieties. For calculations the CAS is used mainly within the static isolated symbolic representation, adding graphic representations for interpreting or explaining symbolic results. The advantage of using CAS is the notation of solutions on a symbolic level, especially while working with equations with parameters. Like in the case of working with functions the communication with the tool is possible in a language close to the traditional mathematical language. The CAS is a consultant in the sense of a formulary for symbolic solutions especially for polynomial equations of order 2 or 3.

We summarize the meaning of CAS concerning the content *equations*:

- CAS are an interactive digital formulary;
- CAS expand the solving of equations to equations usually not considered in school mathematics based on paper and pencil algorithms. They allow symbolic solution e.g. of equations of degree 3 and numeric and graphic solutions of (nearly) any equation.

Especially the following mathematics competencies are supported: Learners

- can interpret solutions of equations and systems of equations, given as a result of a CAS-command, on a numeric, symbolic and graphic level concerning the existence and number of solutions;
- especially interpret CAS-solutions of (complex) equations which they cannot solve by hand;
- see the meaning of symbolic solutions for proofs and argumentations, and the meaning of numeric and graphic solutions especially for real-life-problems.

5 Conclusions

The examples in Sect. 4 show various aspects of a CAS while working with sequences, functions and equations.

- First of all, the CAS is a tool, which allows calculation on a symbolic level in notation or language close to the mathematical language. There are some special tool commands, which have to be learned to use the tool language. But moreover, there are worthwhile and fruitful notations—like seeing parameter dependent functions as functions of several variables—which allow an efficient working style in problem solving processes. A CAS is an isolated representation which may be used statically or dynamically.
- Second, working with a CAS on the symbolic level has to be seen in relation with other representations, especially with numerical and graphical representations (the aspect of multiple representations). These additional representations allow interpretations of symbolic results and expressions.
- Third, the relation to the dynamics of the representations shows especially the dynamic aspect of variables, especially while using multiple representations.
- Finally, the CAS-use depends on the aim of mathematical activities. The CAS is used

- for calculations on a symbolic, numeric and graphic level (Calculate);
- as a consultant in the sense of using a formulary (Consult);
- for controlling calculations on a graphical, numeric or symbolic level (Control);
- for the communication between the user and the media or digital tool, but also for the communication between the user (student) and someone who has to interpret or understand the CAS-solutions (like a teacher) (Communicate);
- for solving problems on a heuristic level or discovering problem solving strategies (Discover).

The developed competence model is a theoretical or normative model. In the first line, the main reason was to get a basis for argumentations for the benefit of using a CAS. Concerning the empirical justification of the theoretical model and with the aim of constructing an *empirical* competence model, some questions have to be answered.

1. The validity of the model: Problems for each "cell" of the topic-representation matrix have to be developed and students' solutions have to be evaluated. Moreover, it will be necessary to concentrate on specific topics—concerning sequences, functions and equations—for students in specific grades.
2. From a qualitative to a quantitative model: The PISA studies use a model with a numerical competence scale, which is based on the relative frequency with which students are able to solve a problem. The problem that has been solved successfully is taken as a measure of the difficulty of the exercise. The scale is standardized on a mean value of 500 with a standard deviation of 100 (OECD, 2003). This might also be an aim in the context of this competence model.
3. Diagnostic: The competence model will also be used for diagnostic reasons to evaluate the "tool-competencies" of students. But diagnostics are only the first step while improving students' competencies in this area. The second step is to establish consequences to *improve* students' competencies. How can a student in the best way be supported to attain a better understanding and higher competencies in working with CAS?

The evaluation of the competence model on the one side and to improve students' abilities in using CAS will be the up-coming challenge.

References

Ainsworth, S. (1999). The functions of multiple representations. *Computers & Education, 33,* 131–152.

Aldon, G., Artigue, M., Bardini, C., & Trouche, L. (Eds.). (2008). *Une étude sur la conception et les usages didactiques d'une nouvelle plate-forme mathématique, potentialité, complexité.* http://educmath.inrp.fr/Educmath/dossier-parutions/experimentation-collaborative-de-laboratoires-mathematiques. Accessed May 29, 2016.

Artigue, M. (2002). Learning mathematics in a CAS environment: The genesis of a reflection about instrumentation and the dialectics between technical and conceptual work. *International Journal of Computers for Mathematical Learning, 7,* 245–274.

Bauer A. (2013). Reasoning with multiple and dynamic representations. In E. Faggione & A. Montone (Eds.), *Proceedings of the ICTMT 11*, Bari. (pp. 327–329). http://www.dm.uniba.it/ictmt11/download/ICTMT11_Proceedings.pdf

Clark-Wilson, A. (2008). *Evaluating TI-Nspire in secondary mathematics classrooms*. Chichester: The Mathematics Centre of University of Chichester. Research report. www.chiuni.ac.uk. Accessed May 29, 2016.

Drijvers, P. et.al. (2010). Integrating technology into mathematics education: Theoretical perspectives. In C. Hoyles & B. Lagrange (Eds.), *Mathematics education and technology-rethinking the terrain. The 17th ICMI study* (pp. 89–132). Dordrecht: Springer.

Drijvers, P., & Weigand, H.-G. (Eds.). (2010). Handheld technology in the mathematics classroom —theory and practice. *ZDM—The International Journal on Mathematics Education,* (7).

Guin, D., Ruthven, K., & Trouche, L. (Eds.). (2005). *The didactical challenge of symbolic calculators*. New York: Springer.

Hoyles, C., & Lagrange, J.-B. (Eds.). (2010). *Mathematics education and technology—Rethinking the terrain. The 17th ICMI Study*. New York: Springer.

Ingelmann, M., & Bruder, R (2007). Appropriate CAS-use in class 7 and 8. In J.-H. Woo et al. (Ed.), *Proceedings of the 31st Conference of the International Group for the Psychology of Mathematics Education*. PME. Seoul: The Korea Society of Educational Studies in Mathematics.

Kieran, C., & Drijvers, P. (2006). The Co-emergence of machine techniques, paper-and-pencil techniques, and theoretical reflection, A study of CAS use in secondary school algebra. *International Journal of Computers for Mathematical Learning, 11*, 205–263.

Kultusministerkonferenz (KMK). (2004). *Bildungsstandards im Fach Mathematik für den mittleren Schulabschluss*. München: Wolters Kluwer.

Kultusministerkonferenz (KMK). (2012). *Bildungsstandards im Fach Mathematik für die Allgemeine Hochschulreife*. München: Wolters Kluwer.

Ng, O.-L. (2016). Comparing calculus communication across static and dynamic environments using a multimodal approach. *Digital Experiences in Mathematics Education*, 1–27.

Niss, M. (2004). Mathematical competencies and the learning of mathematics: The Danish KOM project. In A. Gagtsis & S. Papastavridis (Eds.), *3rd Mediterranean Conference on mathematical education, 3–5 January 2003, Athens, Greece* (pp. 115–124). Athens: The Hellenic mathematical society, 2003.

OECD: Organisation for Economic Co-operation and Development (Ed.). (1999). *Measuring student knowledge and skills. A new framework for assessment*. Paris: OECD Publication Service. http://www.pisa.oecd.org, http://www.pisa.oecd.org/document/58/0,2340,en_32252351_32236159_33688954_1_1_1_1,00.html

OECD: Organisation for Economic Co-operation and Development (Ed.). (2003). *The PISA 2003 Assessment frmework: Mathematics, reading, science and problem solving knowledge and skills*. Paris: OECD Publication Service. http://www.pisa.oecd.org, http://www.pisa.oecd.org/document/29/0,3343,en_32252351_32236173_33694301_1_1_1_1,00.html

Pierce, R., & Stacey, K. (2004). A framework for monitoring progress and planning teaching towards the effective use of computer algebra systems. *International Journal of Computers for Mathematical Learning, 9*, 59–93.

Tonisson, E. (2015). Differences between expected answers and the answers given by computer algebra systems to school equations. *The International Journal for Technology in Mathematics Education, 22*(2), 71–77.

Vollrath, H.-J. (1989). Funktionales Denken. *Journal für Mathematikdidaktik, 10*, 3–37.

Weigand, H.-G. (2008). Teaching with a symbolic calculator in 10th grade—Evaluation of a one year project. *International Journal for Technology in Mathematics Education, 15*(1), 19–32.

Weigand, H.-G. (2010). Book review of C. Hoyles & J.-B. Lagrange (Eds.). (2010). *Mathematics education and technology—Rethinking the terrain. The 17th ICMI Study*, Springer: New York. ZDM—*The International Journal on Mathematics Education 42*(7), 801–808.

Weigand, H.-G., & Bichler, E. (2010a). Towards a competence model for the use of symbolic calculators in mathematics lessons—The case of functions. *ZDM—The International Journal on Mathematics Education, 42*(7), 697–713.

Weigand, H.-G., & Bichler, E. (2010b). Symbolic calculators in mathematics lessons—The case of calculus. *International Journal for Technology in Mathematics Education, 17*(1), 3–15.

Weigand, H.-G., & Bichler, E. (2010c). The long-term project "Integration of symbolic calculator in mathematics lessons"—The case of Calculus. In V. Durand-Guerrier, S. Soury-Lavergne, & F. Arzarello (Eds.), *Proceedings of the Sixth Congress of the European Society for Research in Mathematics Education*. 2009, Lyon (France) (pp. 1191–1200). http://ife.ens-lyon.fr/publications/edition-electronique/cerme6/wg7-15-weigand-bichler.pdf

Weigand, H.-G. (2013). Looking back and flashing ahead—Didactical implications for the use of digital technologies in the next decade. In E. Faggione & A. Montone (Eds.), *Proceedings of the ICTMT 11*, Bari (pp. 298–303). http://www.dm.uniba.it/ictmt11/download/ICTMT11_Proceedings.pdf

Zbiek, R. M., Heid, M. K., Blume, G. W., & Dick, Th P. (2007). Research on technology in mathematics education. In F. K. Lester (Ed.), *Second Handbook of Research* (pp. 1169–1206). Charlottte, NC: Information Age Publishing.

Part IV
New Spaces for Teachers

Innovations Through Institutionalized Infrastructures: The Case of Dimitris, His Students and Constructionist Mathematics

Chronis Kynigos

Abstract The paper discusses the case of Dimitris, a secondary mathematics teacher, who selected three micro-experiments from an institutionalized portal, re-mixed them and then gave his version to his students who in turn made their own changes and constructions. The case is discussed in the frame of the potential for institutionalized portals and digital infrastructures to afford pedagogical innovation which in this particular instance was about designing and re-mixing digital artefacts as an activity for educators, designers, teachers and students alike. Innovation is considered simultaneously at diverse levels, the representational affordances of digital artefacts, the potential for experiential mathematics for students, the potential for teacher-designer expressivity and the potential for economy-of-scale interventions. Dimitris' changes were about the level of abstraction of the available linked representations in a simulation, about restructuration by bringing up front the notion of equivalence in solving equations, about encouraging the use of the negation of a property in a geometrical justification and about laying the ground for students to discover the usefulness of linear functions in working with geometrical properties. The students employed equivalence in a situated context, created an auxiliary point and segment to think around a geometrical property and embedded a linear relationship between segment lengths to create a rectangle which can never be a square. The paper discusses the potential for accredited large-scale institutionalized infrastructures to become the starting point for the generation of personalized living digital artifacts for both teachers and students rather than a showcase of exemplary interactive artifacts.

C. Kynigos (✉)
Educational Technology Lab, School of Philosophy, National Kapodistrian University of Athens and CTI-Diophantus, Athens, Greece
e-mail: kynigos@ppp.uoa.gr

E. Faggiano et al. (eds.), *Innovation and Technology Enhancing Mathematics Education*, Mathematics Education in the Digital Era 9,
DOI 10.1007/978-3-319-61488-5_9

1 Introduction

In their paper in this volume (Chap. From Acorns to Oak Trees: Charting Innovation Within Technology in Mathematics Education) identify multiple representations and the mathematical figure to be the backbone of innovation brought to mathematics education through digital technologies. Mathematical meaning is both generated and expressed in different ways by means of these two affordances. The figure is no longer an object but an instance of a class of objects since no figure in mathematical digital media is expected to be treated as invariable. Representations are no longer confused with mathematical concepts in themselves. When a student co-varies more than one linked representations of a concept, discrimination of the concept itself becomes more natural. So, the nature of innovation brought by digital media to mathematics education is that they afford generation and expression of meaning in fundamentally different ways than pre-technological media. As a corollary, digital simulations afford the possibility to embed mathematical concepts in a vast variety of models of physical or societal phenomena and situations allowing students to appreciate the relevance and role of mathematics in their own reality.

More recently however, different kinds of innovations affecting mathematics education are becoming a real potential, apart from innovations with respect to expressivity. Authoring systems allowing teachers to design their own artefacts and tasks for their students, embedding content of their own choice. Systems providing structured and searchable access to large quantities of resources from anywhere by means of any computational device at any time. Systems supporting social engagement, argumentation and sharing of artefacts. Videos and e-books as new kinds of mediating informational or instructional resources and exercises. Even interactive e-books, such as 'the c-book' (built through the European 'M C Squared' project, Kynigos, 2015), where text and narrative seamlessly mesh with a diversity of dynamic digital artefacts.

In research, there is often a focus on one innovation from the list above, conveying a feeling of fragmentation in the way innovations are addressed. Large portals are seldom designed to focus on a clear pedagogical innovation such as mathematical meaning-making and expressivity. The attention is on quantity, diversity and search-ability. Artefacts designed to afford meaning making are seldom considered by their creators as one amongst different types of media and activities made available to students. The attention is on the affordances of the respective medium and mathematical meaning making is perceived through this medium in a kind of 'silo' approach. This paper takes a flipped stance towards this issue in order to discuss innovation made possible by connecting different types of innovation. It reflects on the potential for an integrated approach to innovation through the description of a case of a teacher and his students engaging in activity based on systemically enabled expressivity and meaning making. Can a teacher and his students use an institutionally accredited portal infrastructure to re-mix digital media in order to construct mathematical meaning? The following two sections provide some contextual framing necessary to discuss the case of Dimitris and his students.

2 The Case of Constructionist Mathematics

Many years ago the innovative role of digital technologies for learning mathematics was primarily perceived as that of an expressive medium for students to generate mathematical meanings by means of programming a computer with a language affording formalism and connected graphical outputs to executed commands and procedures (Abelson & DiSessa, 1981; Papert, 1972). The key idea characterising students' mathematical activity was termed 'constructionism' by Papert and his group (Kafai & Resnick, 1996). By constructing, tinkering with and re-mixing graphical models, which were then the only affordance available, students would have much more dense opportunity to construct meaning than just express meaning with pencil and paper. The process of generating mathematical meaning was then described as containing four sides to it, using ideas, discriminating mathematical essence or structure amongst them, generalizing within the computational context and synthesizing with the same ideas in a more abstract way (Hoyles & Noss, 1987). Some recent versions of such media afford expressing such meanings by means of linked representations, dynamic manipulation and feedback (see, e.g., Zantzos & Kynigos, 2012). Meanings are thus expressed as they are generated and so become visible artefacts amenable to inspection, sharing, discussion and argumentation.

Many things changed since those early days. The advent of Computer Algebra Systems and Dynamic Geometry software afforded representations much closer to the ones traditionally established in mathematics (Artigue, 2002). Dynamic manipulation of these representations was perceived as conveying mathematical meaning and as a new semiotic system in itself (Arzarello et al., 2002; Bussi & Mariotti, 2008; Morgan & Kynigos, 2014). During the era of multimedia in the late 90s attention turned to the agenda of experimentations with simulations by-passing formal symbolic programming, by means of mathematical representations of the underlying simulation rules (Kaput et al., 2002).

The internet and the social web brought attention to argumentation, collaborative learning, publicising and sharing constructs and links with some work on collaborative activity for mathematical argumentation (Stahl et al., 2010) but not so much emphasis on the use of digital media as media to express and construct mathematical meanings. Now, we are in the midst of the era of large portals, spaces with zillions of digital artefacts with searchable meta-data and a great emphasis on video artefacts (Kahn Academy) and e-books. So, what happened to the idea of using digital technology for generating and expressing mathematical meaning? What happened to the idea of how important it is for this media to afford deep structural access (DiSessa, 2001) to its functionality in order to invite mathematical meaning making?

This chapter suggests that it is important to re-consider constructionism as an innovative activity, rich in opportunity for meaning making in the era of large portals and the social web, especially now that expressing meaning, using formalism and coding (the new term for programming) are perceived as central for expressivity and meaning generation. Different types of innovations have attracted attention at different times. How can we approach the challenge of maintaining focus on more than one, on merging the agendas of different stakeholders and forging connections between diverse kinds of affordances?

Let us briefly consider possible meaning-making and expressivity innovations through the affordances of large portals. The most pertinent initiative has not come directly from mathematics education but rather from communities perceiving programming as an expressive medium. For around ten years now, 'Scratch' has been paving the way for a number of recently emerging infrastructures for constructionist activity such as 'Second life', 'Kodu' or 'Minecraft', addressing mostly informal settings and gaining popularity based on their own merit. At the same time mathematics education infrastructures for access to digital media in institutionalized settings such as ministries of education and organizations working on their behalf have grown but mostly seem to be indifferent to supporting coherent pedagogical reform and innovation, apart from a very few exceptions, see for example Benton et al., 2016). The agenda there is mostly to accrue the largest possible volume of resources provided they have a basic accreditation mainly with respect to the validity of and the rights to the embedded information. Very often, these portals give mixed messages with respect to the role and uses of digital artifacts, emphasizing video narratives, links to encyclopedic information, tightly defined exercises usually to be resolved by multiple choice questions and finally some simulations affording a simple experiment.

So, this chapter raises the question of how might it be made possible to embed the potential of supporting mathematical meaning making and expressive activity in these institutionalized infrastructures without disputing their agendas to democratize access to information and easily understood affordances but at the same time seeding affordances for pedagogical innovation. The pedagogical innovation considered here involves educators, teachers, designers and students making structural changes to artefacts made widely available through such institutionally accredited portals. To use Chevallard's metaphor, the chapter considers the question of whether it is possible to go to a human knowledge exhibition centre and place some exhibits allowing visitors to engage in producing their own knowledge (Chevallard, 2012).

3 The Case of Institutionalized Innovation

I take the case of the 'Digital School' infrastructure of the Greek Ministry of education created in the past four years and in particular two co-existing portals, the 'Interactive books' portal and the 'Photodendro' portal at http://photodentro.edu.gr/aggregator/?lang=en. The former contains the original unique curriculum books (in Greece there is only one institutionally accredited curriculum book per subject) enriched by the inclusion of links to a variety of artefacts in amongst the text and in tight relation to it. The latter, Photodendro, is a classic portal with carefully organized meta-data for each of the artefacts which began by containing the artıfacts of the interactive books portal and has been growing since, but with no connection to the curriculum books.

The challenge to embed constructionist mathematical meaning making in the affordances of institutionalized infrastructures is complex and the argument here is that it may be feasible and will certainly constitute and generate innovative ways and new potential to proliferate meaning making uses of digital media and the pedagogies necessary for that to happen. Ruthven discusses some important parameters facing the task for teachers to sustainably infuse meaning making pedagogies based on the use of digital expressive media, such as accreditation, curriculum script, time economy (Ruthven, 2009).

My role in the design and development of this infrastructure was to coordinate the design and development of artefacts for the domain of mathematics from year 3 through 11. I worked with a team of 30 professionals selected so as to have diverse expertise, comprising of technical knowledge, pedagogical design knowledge and mathematical knowledge (Fischer, 2012; Kynigos, 2007). In a course of four years, an impressive number of 1800 original artefacts were developed and uploaded in the two portals, almost all of them constituting what we call 'micro-experiments', i.e. tightly focused microworlds, objects with which students can experiment and dynamically manipulate some simulation or problem embedding mathematical concepts in order to discuss and answer in the classroom a set of closed questions and occasionally a final open question involving invitation to some constructionist activity (Kynigos, 2012). To orchestrate a sound integration of the knowledge and experience inherent in the design team and also to create the conditions for creative designs, I made sure that each of artefacts was created by 2–3 designers with diverse expertise and internally reviewed by another team member in a way visible to all designers (for a discussion, see Clinton & Hokanson, 2012; Gero, 2010). The underlying authoring tools were Geogebra, MaLT—a 3D web version of e-slate Turtleworlds, (Zantzos & Kynigos, 2012; Kynigos, 2004)—and some custom widgets built with flash and other tools.

The question I'd like to discuss in this paper is how may it become possible for this kind of infrastructure to support constructionist activity. I thus asked Dimitris, a mathematics teacher with a masters degree in mathematics education and some (but not extensive) experience with constructionist media, to pick a small number of artefacts, change them and then give them to his students to engage in mathematical activity.

The paper revolves around three examples of Dimitris' work (see Kynigos, 2007; Kynigos and Diamantidis, 2014 for a background discussion). The corresponding artefacts will thus be discussed with respect the way they were originally designed and the way they may be used by people who play different roles, such as the role of the designer, the educator and the student (Kynigos, 2002). The main perspective and pursuit is to talk about the fact that Photodentro and the interactive books portal can play the role of the resource, of the available infrastructure, but it can also play the role of a springboard for design, creation and development for all the people involved in education, from the educator to the student (Pepin et al., 2013; Gueudet & Trouche, 2012).

3.1 The Context of Large Scale Initiatives

Firstly, for Mathematics we had the possibility and the opportunity from very early on to work in conjunction with other major initiatives of the Ministry that happened to take place around the same time. One of those initiatives was the committee for the new curriculum, which for the first time placed great emphasis on mathematical activity, that is what is proposed for students to do in order to become personally engaged with the concepts of mathematics by actually utilizing them and have a mathematical experience. Additionally, great emphasis was been placed in this new curriculum on mathematical literacy, that is the approach to mathematics as an important societal asset, a cultural tool concerning everyone's actual life and not the mere abstract end-product of a scientific field. The second major concurrent initiative of the Ministry was what was named 'second level Training program', which supports in-service teachers to make use of the two portals and all the infrastructure to a very large degree, focusing on mathematics education, which is the blending of new and established methods, techniques and teaching practices http://b-epipedo2.cti.gr/. Thus, the 'Digital School' initiative was for us the third field of contribution and intervention, the one that constitutes the infrastructure in terms of available material.

Beyond any doubt there are great many matters that are addressed with these infrastructures, and thus the core part of their designers' agenda, such as the availability of textbooks or resources online for everyone, from anywhere and at any time. What we were interested in the domain of mathematics was to look for added pedagogical value that may be involved in utilizing them, what the educator or the student can do that would otherwise be very difficult to be done without these technologies. This was our main focus for the subject of mathematics. Reinforcing the possibility for students to have a personal experience of mathematical reasoning in situations that are realistic for them, by utilizing the available infrastructures as the tool for expressing concepts of mathematics by using them in the context of mathematical literacy. The interweaving between the three major actions was especially important for us as to eliminate the confusion that is often caused by the feeling of fragmentation between intervention actions in the field of education.

3.2 The Quest for Added Pedagogical Value

We addressed the challenge for the portal to afford added pedagogical value not only with respect to student activity but also with respect to the potential for teacher designs. In the paper there are three examples, of three artefacts, that incorporate different technologies and concepts of mathematics and I will describe the ways that they have been constructed by the original micro-experiment designers, by Dimitris as a teacher re-mixing these accredited designs and by his students for each of them to express concepts of mathematics.

We perceived the added value for students to be that they can utilize these infrastructures to strengthen their mathematical experience and the feeling that mathematics is something that is realistic, interesting, fun as well as beautiful and mainly that it's something useful for our everyday lives. Apart from the students, we also aimed at affording added value for the teacher in the role of a professional designer. We saw the potential for the portal to be that of a springboard for the design and development of artefacts, not only by specialists and researchers but also by the educators themselves. In the original team of mathematics designers of the portal artifacts, we were 35 specialized colleagues and all the artefacts of mathematics in the Photodentro, around 1800 artifacts, were original and developed by the group. This team of colleagues consisted of technicians and educators that had overall knowledge of pedagogy, mathematics and technology. Not even a single artefact was developed without having all these three aspects of know-how underlying, by the people that made it. Beyond this highly specialized team, this infrastructure also provides the educator himself with the possibility of modifying his own artefacts, design activities for his students and his own repertoire can be contained in these activities, his own "suitcase" or "la valise" -as the French call it, see Pepin et al. 2013—of digital artefacts that are modifications of the ones that can be found at the Photodentro. And these possibilities also exist for trainers and councilors for them to strengthen and create seminars of training and practice between colleagues who discuss matters of didactics.

3.3 Emphasizing Equivalence to Solve an Equation: Dimitris' Re-Mix of a 'scales' Simulation

This section contains a discussion of the first example involving the representations in a simulation of a physical phenomenon embedding the concept of an equation and the distinction between equation and equivalence. Dimitris chose a classic equation problem embedded in a scales task residing in the 8th grade interactive textbook (and also in the Photodendro portal). In the paper version the task refers to a picture of the scales. In the interactive book, it is dynamically manipulable, affording the adding or taking away of known and unknown weights and the simulation of a balanced or tilted scale (Fig. 1). The weights are represented iconically with a different shape and color for the unknown which is also marked with the letter 'x'. The 'known' weights of 200 grams have the respective number marked on one of the side facing the user. So, the idea is that students will experiment and come up with a solution involving the isolation of one unknown weight on one scale and the number of known weights on the other, needed to balance the scale. They are supposed to connect this with the process of solving an equation. The digital tools for experimentation available in this simulation are sliders that change the number of known and unknown weights on each one of the

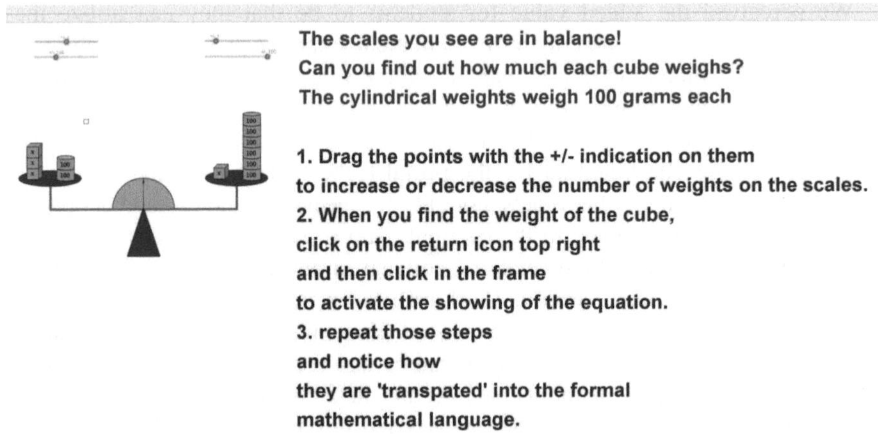

Fig. 1 The "scales" micro-experiment

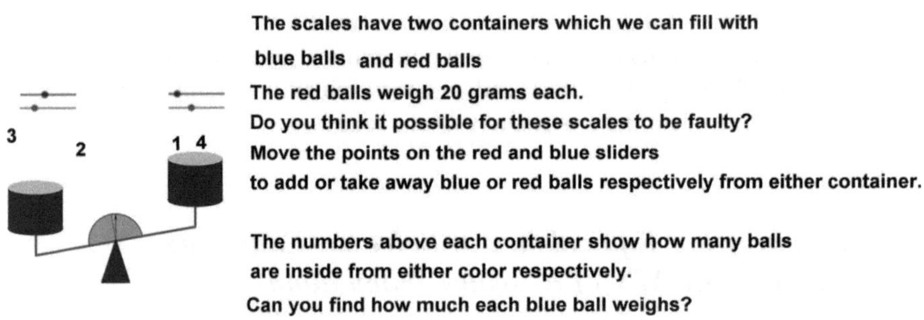

Fig. 2 Dimitris' "scales" experiment

two sides of the scale. Dragging a slider dynamically adds or subtracts weight icons on the simulation.

Dimitris however had other ideas regarding how to use it in order to make the mathematics more interesting for his students and changed the artefact completely, he created a modified version (Fig. 2). In his version of the simulation, the semantics for the weights and their measure are not iconic, the user imagines the weights inside two pots (for a discussion of artefacts as representations, see Morgan & Kynigos, 2014). The semantics are numerals for the number of weights and color for known (red) and unknown (blue) weights, all changeable by means of respective sliders. Dragging a slider now changes the digit of the respective type of weight. Up till now, the simulation appears to be the same just with different, more abstract, representations. Things change dramatically however by a textual question asked to the student: "is there any chance that the scale is faulty"? Dimitris wanted to get the students to work that one out and then find the value of the unknown weight too.

Let's see now how this artefact was used by the students. One way that the children themselves thought of how to deal with this was to put the same number and combination of weights on each side, despite the fact that the blue weight was unknown. The scale turned out to have a visible tilt. They then added known weights until it apparently balanced to find out the extent of the fault. In this case, their thinking dealt with the unknown value as an object which is at the heart of mathematical thinking. It means that they could cope with imagining that the number had been found and additionally accepting that the scale is broken and there is no balance. So, the students found that the scale had an error of 40 kg since it was balanced when two red weights of 20 kg each were added to the equivalent weights. Afterwards, in their effort to find how much the unknown blue weights weighed, a strategy that a group of kids thought of was to put the scale in balance and then increase the number of blue weights by one and then see how many reds they needed to add in order for it to become balanced again and in that way figure out the solution. Dimitris' agenda was for the students to see the value of equivalence and appreciate that it lies behind the concept of equation. The students' changes allowed them to be able to use a faulty balance to work out their equation problem and estimate the extent of the fault. Although maybe in this case they did not change the functionality of the simulation, they did understand it to an extent allowing them to resolve a problem.

The original scale micro-experiment was designed and developed in a way which was typical of the process of designing all of the original artifacts for the two portals. Special care was taken so that each one emerged out of the collaboration of 2–3 designers with diverse expertise, at least one with technical know-how and one with pedagogical design experience. First and foremost the directive to the designers was to pay special attention to the correctness of the mathematics embedded in the artifact in essence and in the way is was conveyed by means of its representation. The agenda here was for the portal to be widely accepted and compatible with the systemic agenda of accreditation. Equivalently, the developers were asked to clearly negotiate the pedagogical agenda for its use, i.e. to anticipate what the students would do with it as an expressive tool. Thus, although the development of the two portals was funded simply to play the role of a resource, for the domain of mathematics it was almost exclusively designed as a set of artefacts for the students to do mathematics with, i.e. to experiment, to modify, to think around and to justify behaviors, properties and the changes they made.

3.4 Constructionism for All in a Classic Geometrical Problem

The second example shows how students made structural changes to an artefact already changed by Dimitris. We now have a classic geometrical example, which is in the interactive textbook (Fig. 3), two concentric circles are given, with two corresponding diameters and students are asked to tell what kind of quadrilateral is formed if the four intersection points of the diameters and the circles are connected and also justify their answer. Next, they are asked, by increasing and decreasing the lengths of the diameters (and thus the circle size) and changing their direction to make various quadrilaterals and justify their constructions (e.g. rectangle, rhombus etc.). But Dimitris wanted more in terms of the possibility of engaging his students in mathematical thinking. So, he made two line segments, where one of them was a diameter and the other was a mere chord of an arc whose size could be modified. He asked his students if a parallelogram is formed, as well as when and why. He asked them to explain what is happening on the shape and construct different parallelograms of their own choice. We can see that the questions are more open, they invite students to create quadrilaterals and to explain how a figure does not belong to a particular class when the properties necessary for it to do so do not apply (Fig. 3).

Now let's look at some students' activity. A certain student thought of connecting the two midpoints, one of the diameter and the other of the chord, O and O', and observe what happens to the OO' segment in order to provide his explanation. This is especially important, since the student felt that he had the right and that it is part of his role to tamper with the software, to add a line segment which will help him think and then dynamically manipulate it.

It is also important that in order for such a thing to happen, the teacher needs to create and encourage this norm in the classroom, i.e. that these tools are tools for

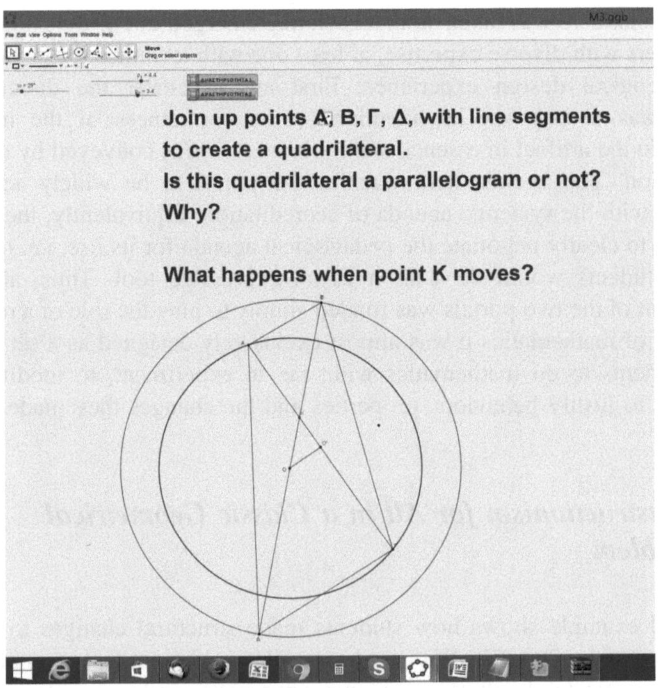

Fig. 3 Dimitris' geometry experiment

experimentation and engagement as well as modification and tampering. In support of this, let us look at the things the children were saying while using the tool. We have a student who says "this quadrilateral is not a parallelogram since the opposite sides are not parallel and equal". So far she answers the question. But, without being asked, she goes on to say, "if the point K is moved along the diameter that is there, the length of the chord, from which this point passes through, will change. So, as K comes closer to D, the length of the side BG increases and of GD decreases". This conclusion was not part of the question. It is however a mathematical formulation, a conjecture and a topic for thought in class, that this particular student felt it is a valid thing to do, it's within the norm, within reason to do so. Despite the fact that there was a specific question, children felt free to think of the mathematics around the question and not just answer it and move on to the next matter and the next problem.

These digital infrastructures, are thus infrastructures that encourage and allow children to express themselves, to attribute meaning in what they're doing. They include interdependent representations as is shown in the first example, which is a very important matter in mathematics, they help children become detached from the representation and understand that there are concepts behind it. They allow dynamic manipulation, which is a new way to represent concepts in mathematics as shown in the second example, a way that was not available before. They allow teachers and students alike to have deep access to functionalities. In the first two examples, Dimitris modified the scale simulation and took away a property of one segment (diameter) to drastically change a mathematical problem. The students respectively changed the functionality of the experiment by employing equivalence to resolve an equation and by adding a segment to help them study a property. It is important to point out that, the systemic agenda was not challenged, i.e. the micro-experiments were in themselves accredited resources aiming at a wide-scale use with not much emphasis on pre-requisites for a high teacher T.P.a.C.K. level or students highly tuned to experimentation with digital media. However, their affordances allowed re-mixing from teacher and students in a large variety of ways. The former example involved some use of multimedia objects and the representation of a physical phenomenon where mathematics was embedded, the latter was a simulation of a mathematical object who's abstract generalized nature could be perceived by means of dynamic manipulation of the representation.

3.5 Employing Algebra for a Geometrical Construction

The third example is about how we can combine concepts of mathematics that lie in different sections in the curriculum, to such an extent so that they are mistakenly considered to be unconnected. It is also about how students can for themselves invent ways to use mathematical concepts. The respective artefact from the digital school is for the 7th grade and it is about the rhombus and the square, apparently geometry again. This artefact incorporates mathematical expression through computer programming, it is made with a new web-based 3D version of E-slate Turtleworlds which we call ‘Turtlesphere’ which we developed at the Educational Technology Lab (http://etl.ppp.uoa.gr/malt2).

Students are put in the position of an engineer, and are asked to de-bug a model that one of their fellow students has apparently made and that is not working correctly. “Yiannis’ team”, we read in the problem text (Fig. 4), “tried to make a procedure to construct a square, without success, can you help Yannis fix the procedure so that when it is executed a square is always constructed”? Students can experiment with the sliders by altering the variable values dynamically, look at the code to figure out what change is needed, which property of the square is missing in this procedure. What is missing here is that the angles should be 90 degrees rather than variable values independent from each other. They get into the formalism of mathematics, correct the code, run the procedure and observe if it works or not.

First, let us discuss Dimitris’ modification (Fig. 5, first column). He turned this problem to be about parallelograms providing one that doesn’t work in order for the students to experiment and find out, think of what they need to do to make this shape become correct once again. After a lot of discussion, the children managed to realize that the subsequent angles of the turns are supplementary and that this suffices for the parallelogram to be fixed and nothing else is required.

Dimitris then gave them another problem, he asked if they can make a rectangle, that can never be a square. The students came up with various strategies. For example, they put a variable x for one of the two opposite sides and they used $x + 20$ in one case and $2 \times x$ in another, for the other side (Fig. 5, columns 2 and 3). What did the children do here? In order to express a generalized inequality in the mode, without anyone telling them, they thought of incorporating the linear function as an element of changing and modifying the model. With this artefact the

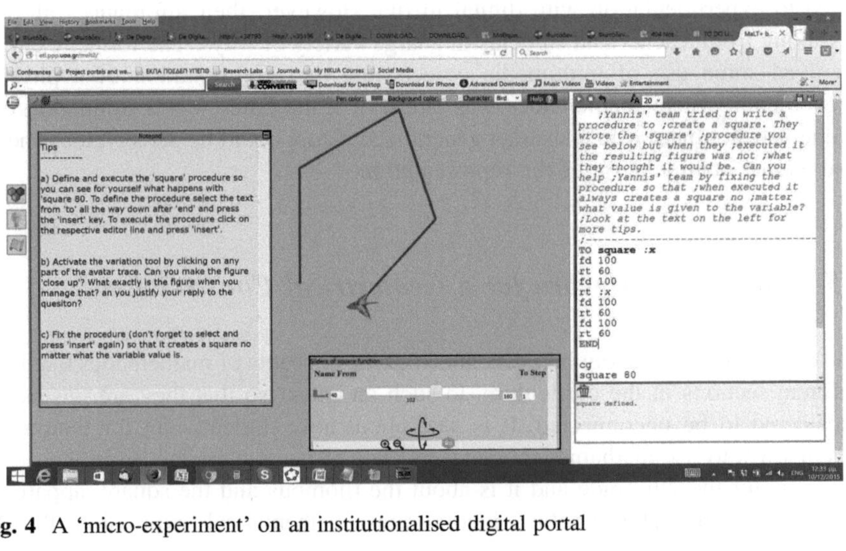

Fig. 4 A ‘micro-experiment’ on an institutionalised digital portal

To parellelogram :*x* :*y*	To change :*x*	To myrectangle :*x*
Fd50	fd :*x*+20	Fd :*x*
Rt30	rt 90	Rt 90
Fd :*y*	fd :*x*	Fd :*x**2
Rt :*x*	rt 90	Rt 90
Fd50	fd :*x*+20	Fd :*x*
Rt30	rt 90	Rt 90
Fd 100	fd :*x*	Fd :*x**2
Rt150	Rt 90	Rt 90
end	end	end

Fig. 5 Students' constructions as responses to Dimitris' task

students themselves came up with an algebraic solution to a geometrical problem and incorporated a linear relationship between two generic numbers, the length of the opposite sides.

4 Discussion

These three examples provide a flipped perspective on innovation. Rather than addressing one type of innovation such as a special artefact or infrastructural affordance, the chapter considers innovation in an integrated pragmatic approach through a teacher and his students' activities with media residing in institutionalized infrastructures. How can we as researchers elaborate cases of the whole cycle of infusing an agenda for added pedagogical value into a systemic agenda for generating an accredited institutionalized digital infrastructure in an education system? In this chapter, the cases are meant to help identify the challenges and the potential rather than to convey showcase exemplary stories. My participatory involvement in the process and the reflections expressed in this chapter led me to identity the following three issues.

1. Considering the role of constructionist mathematical activity within an evolution of broader agendas for educational innovation.
2. The need to address the essence and characteristics of differing but not necessarily conflicting agendas so as to generate and construe integrations where possible.
3. Reflecting on the potential to integrate pedagogical added-value innovation such as constructionist mathematical meaning making and expressivity, with larger equity, political and large-scale economy agendas.

The first issue to consider is the role of constructionist mathematics with expressive media as attention to technological infrastructures, societal norms and political agendas evolve in time. In the short history from the advent of digital media, this innovation has been highlighted, underplayed, challenged and considered obsolete or wanting with respect to original perceived promises. In this era of social media and large portals, how can systemic agendas co-exist and integrate with an agenda for meaning making with constructionist expressive media? Scratch is a case where the two are integrated but with little as yet emphasis on mathematics albeit some focus which had been given recently (Benton et al., 2016). The Photodendro mathematics portal can in the same sense be considered as an effort to integrate portal agendas with constructionist mathematics. It may be perceived as a resource ready to be used in traditional classrooms by simply having teachers support students to follow the tasks given by the micro-experiment resource. Even at that level, it is designed to widely convey the idea that mathematics is not an abstract isolated field of study which for some unexplained reason is good for the future citizen. Instead, that it may be connected to experimental activity, to engagement with something which is relevant and interesting and most importantly, where sharing and exposing the process of mathematical thinking by means of sharing thoughts, ideas and actions on digital representations and their manipulations is recognized, explicit and valued in mathematics courses.

Let's consider the three micro-experiments as they stood in Photodendro originally. The scales simulation can be used for experimenting with iconic representations of physical objects connected with mathematical properties and dependencies. Geometrical figures can be manipulated dynamically to generate understandings of embedded properties. The mathematical formalism which created figural models can be inspected and changed so as to embed a property necessary to construct a mathematical figure. This paper argues that the same infrastructure may invite going much further with respect to added value agendas. It can also be perceived as a stepping stone for teacher engagement with design like in the case of Dimitris and constructionist mathematics for the student as in the case of the second and third example given in this chapter.

The same scales micro-experiment can be re-mixed so that it lends itself to experimentation with mathematical representations and a focus on equivalence rather than equations, geometrical figures can be manipulated dynamically to express conjectures regarding the consequences of inherent properties, rectangles can be given custom properties expressed algebraically to achieve idiosyncratic behaviors. These thoughts only help raise the issue which in any case remains, how to think of constructionist mathematics not as a silo pedagogy or activity but as an element amongst the affordances of digital infrastructures design for broader reforms focusing on e.g. equity, scale economy, digital citizenship etc. Innovations thought of as silo interventions have little chance of being sustained and proliferated.

The second issue refers to the agendas themselves. Are societal (social web tools) or systemic (portal) agendas necessarily detrimental to the innovation of constructionist meaning making in mathematics? This chapter makes the case that this is not so, leaving an open issue to identify and operationalise agenda integrations, embedding innovative activity in equity and economy-of-scale interventions. The strategy with the Photodendro for mathematics was that of focusing on added pedagogical value rather than creating a portal which would be a centre for exhibiting any type of resource. This took some discussion amongst the original 30 design professionals. Enabling the re-mixing of their own constructs was not readily accepted or understood by all. Neither was conforming to a coherent pedagogical framework. The idea of a pedagogical framework was not clear in any case, some teachers initially aiming to just upload coaching exercises they had already created. My task and challenge was to explain the constructionist agenda in arguments understood and accepted by those colleagues who were much more representative of the average mathematics teacher. This is a much harder task than it looks, and experience with the ways in which digital media has been used in classrooms worldwide is full of stories of why this is so (Ruthven, 2009). Furthermore, embedding innovative activity in one amongst diverse large scale initiatives has thin chances to survive, initiatives emerge from different needs and are often concurrent as was the case with the new Greek curriculum for mathematics and the large scale in-service education program. It would be very normal for the curriculum to undermine this kind of mathematical activity had it not been a lucky circumstance that there was a concurrent curriculum reform involving like-minded professionals. Another aspect of the strategy for mathematics was for the essence of what was perceived as added value which was to include and democratize engagement for all parties concerned, educational designers, teachers in a designer role, students in both a designer role and in engagement with mathematical activity. The innovation strategy thus for the mathematics portal was driven by an emphasis on added pedagogical value based on the importance of experiential mathematics and the democratization of resource design and re-mixing.

The third issue is on what is in it for mathematics education. That is, beyond the inevitable defensive argument that the value of mathematical thinking may survive an era where the focus is on information, administration, collective engagement, social pluralism and economy of scale, what are the arguments for how it may actually gain from added value strategies? In the chapter we discussed some examples, where mathematical representations and figures were in focus. Dynamic manipulation and engagement with dynamically linked representations may gain much higher visibility and accessibility, especially when, as in the case of Dimitris' scales re-mix, switching from one type of representation to another is made so seamlessly. Allowing for different conceptual fields (to use Vergnaud's term, 2009) to emerge placing for instance geometrical properties with functions in the same kernel may generate many more opportunities for mathematical meaning making as in the rectangle failing to ever become a square built by Dimitris' students. Wilensky and Papert (2010) referred to this as 'restructurations'. At another level, recognizing and supporting the design aspect of the teaching profession widely maybe afforded in new ways, think of Dimitris for instance putting together his own 'valise' (to use Pepin et al's term, 2013) of resources continually re-mixing and tinkering with each one and using this experience to reflect on his own pedagogy in TPD courses.

So this is a way in which institutionalized digital infrastructures could afford added pedagogical value for students and also for teachers, designers and teacher educators and consultants. Where is the innovation? it is inherent in the whole of the above phrase. In that the teacher can be assisted by these infrastructures in order for his practice to acquire a higher element of orchestration by means of using them to generate norms for mathematical literacy in the classroom, as shown by the natural way Dimitris' student articulated a mathematical argument during the concentric circles investigation. In that the portal can also be used by teachers to engage in their own research and professional reflection. In that the design element in the teaching profession can also be enhanced in interesting ways given that these digital media can be used as expressive media for design. Teachers can create their own artefacts by remixing the ones given in the portal like Dimitris and thus act as a members of communities that discuss and share such activities and such materials. Consequently, the artefacts now become springboards for student constructions, for design, for creation of such artefacts by the educator and for engagement in communities of educators.

Acknowledgements The research leading to these results has received funding from the European Union Seventh Framework Programme (FP7/2007-2013) under grant agreement no. 610467—project "M C Squared", http://mc2-project.eu. The c-book technology is based on the widely used Freudenthal Institute's DME portal and is being developed by a consortium of nine partner organisations, led by CTI&Press "Diophantus".

This work has been supported by the Greek National project "Digital School Platform, Interactive Books, and Learning Object Repository" (Contract No 296441/2010-2015) that is co-financed by the European Union (ESF) and National funds in the context of Operational Programme "Education and Lifelong Learning" of the Greek National Strategic Reference Framework (NSRF), and is being implemented by CTI "Diophantus".

References

Abelson, H., & DiSessa, A. (1981). *Turtle geometry: The computer as a medium for exploring mathematics*. Cambridge, MA: MIT Press.

Artigue, M. (2002). Learning mathematics in a CAS environment: The genesis of a reflection about instrumentation and the dialectics between technical and conceptual work. *International Journal of Computers for Mathematical Learning, 7*(3), 245–274.

Arzarello, F., Olivero, F., Paola, D., & Robutti, O. (2002). A cognitive analysis of dragging practices in Cabri environments. *Zentralblatt für Didaktik der Mathematik, 34*(3), 66–72.

Bussi, M. G. B., & Mariotti, M. A. (2008). Semiotic mediation in the mathematics classroom: Artifacts and signs after a Vygotskian perspective. In L. English, M. G. B. Bussi, G. Jones, R. Lesh, & D. Tirosh (Eds.), *Handbook of international research in mathematics education*, second revised edition (pp. 746–805). Mahwah: Lawrence Erlbaum.

Benton, L., Hoyles, C., Kalas, I., & Noss, R. (2016) Building mathematical knowledge with programming: insights from the ScratchMaths project. In *Proceedings of the Fourth Constructionism Conference*, Bangok, Thailand.

Chevallard, Y. (2012). Teaching mathematics in tomorrow's society: A case for an oncoming counterparadigm. In *Plenary lecture at 12th International Congress on Mathematical Education*, Seoul, Korea.

Clinton, G., & Hokanson, B. (2012). Creativity in the training and practice of instructional designers: the Design/Creativity Loops model. *Educational Technology Research and Development, 60,* 111–130. doi:10.1007/s11423-011-9216-3

DiSessa, A. A. (2001). *Changing minds: Computers, learning, and literacy.* MIT Press.

Fischer, G. (2012). Meta-design: Empowering all stakeholders as Co-designers. In R. Luckin, P. Goodyear, B. Grabowski, S. Puntambeker, J. Underwood, & N. Winters (Eds.), *Handbook on Design in Educational Research.* Routledge.

Gueudet, G., & Trouche, L. (2012). Communities, documents and professional geneses: Interrelated stories. In G. Gueudet, B. Pepin & L. Trouche (Eds.), *Mathematics curriculum material and teacher documentation: From textbooks to lived resources* (pp. 305–322). New York: Springer.

Hoyles, C., & Noss, R. (1987). Seeing what matters: Developing an understanding of the concept of parallelogram through a logo microworld. In J. Bergeron, N. Herscovics, & C. Kieran (Eds.), *Proceedings of the 11th Conference of the International Group for the Psychology of Mathematics Education,* Montreal, Canada (Vol. 2, pp. 17–24).

Kafai, Y., & Resnick, M. (Eds.). (1996). *Constructionism in practice. Designing, thinking and learning in a digital world.* Wahwah, NJ: Lawrence Earlbaum Associates.

Kaput, J., Noss, R., & Hoyles, C. (2002). Developing new notations for a learnable mathematics in the computational era. In L. English (Ed.), *Handbook of international research in mathematics education* (pp. 51–75). Hillsdale, NJ, USA: Lawrence Erlbaum.

Kynigos C. (2002). Generating cultures for mathematical microworld development in a multi-organisational context. *Journal of Educational Computing Research,* Baywood Publishing Co. Inc. (1 and 2), 183–209.

Kynigos, C. (2004). A black and white box approach to user empowerment with component computing, *Interactive Learning Environments,* Carfax Pubs, Taylor and Francis Group, *12*(1–2), 27–71.

Kynigos, C. (2007) Half-baked microworlds in use in challenging teacher educators' knowing. *International Journal of Computers for Mathematical Learning. 12*(2), 87–111 (Kluwer Academic Publishers, Netherlands).

Kynigos, C. (2012). Niches for Constructionism: Forging connections for practice and theory. In C. Kynigos, J. Clayson, & N. Yiannoutsou (Eds.), *Proceedings of the "Constructionism 2012" International Conference,* Athens, (pp. 40–51).

Kynigos, C., & Diamantidis, D. (2014). The design principles of micro-experiments in the Interactive Digital Student Books' Portal and their role as a basis for developing new artefacts. In C. Skoumpourdi & M. Skoumios (Eds.), *Proceedings of the 1st Conference: The Development of Educational Material for Mathematics and Science* (pp. 823–842). Greece: Rhodes.

Kynigos, C. (2015). Designing constructionist e-books: New mediations for creative mathematical thinking? *Constructivist Foundations, 10*(3), 305–313.

Morgan, C., & Kynigos. C. (2014) Digital artefacts as representations: forging connections between a constructionist and a social semiotic perspective. Special Issue in digital representations in mathematics education: Conceptualizing the role of context and networking theories. In J. B. Lagrange & C. Kynigos (Eds.), *Educational Studies in Mathematics,* Springer Science + Business Media, Dordrecht (Vol. 85, No. 3, pp. 357–379).

Papert, S. (1972). Teaching children to be mathematicians versus teaching about mathematics. *Journal of Mathematics in Science and Technology, 31,* 249–262.

Pepin, B., Gueudet, G., & Trouche, L. (Eds.). (2013). Re-sourcing teacher work and interaction: New perspectives on resource design, use and teacher collaboration. Special issue of ZDM, *The International Journal on Mathematics Education, 45*(7), 929–944.

Ruthven, K. (2009). Towards a naturalistic conceptualisation of technology integration in classroom practice: The example of school mathematics. *Education & Didactique, 3*(1), 131–149.

Stahl, G., Çakir, M. P., Weimar, S., Weusijana, B. K., & Ou, J. X. (2010). Enhancing mathematical communication for virtual math teams. *Acta Didactica Napocensia. 3*(2), 101–114. Web: http://GerryStahl.net/pub/adn2010.pdf

Vergnaud, G. (2009). The theory of conceptual fields. *Human Development, 52,* 83–94. doi:10.1159/000202727

Wilensky, U., & Papert, S. (2010). Restructurations: Reformulations of knowledge disciplines through new representational forms. In *Proceedings of the Constructionism 2010 Conference*, Paris, France, 2010, (p. 97).

Zantzos I., & Kynigos C. (2012) Differential approximation of a cylindrical helix by secondary school students. In *Proceedings of the "Constructionism 2012" International Conference*. Kynigos, Clayson, & Yiannoutsou (Eds.), Athens, 136–145.

Studying the Practice of High School Mathematics Teachers in a Single Computer Setting

Michal Tabach and Galit Slutzky

Abstract Many studies have examined the teaching of mathematics in technological environments that are accessible both to the teacher and to the students. Nevertheless, some classrooms are equipped with only one computer and a data projector. This study examined case studies of four different teachers who had previously worked in the high-tech industry and then became high school mathematics teachers that used technology in the classroom. Two technological environments were examined: (1) an environment in which teachers used a computer and a projector and (2) an environment that also included an interactive whiteboard (IWB). The study aimed at characterizing teaching practices and teacher knowledge in these two environments. An innovative framework was developed, based on three lenses: (1) the teachers' goals; (2) the technological resources used; and (3) the way these resources were used. Findings indicate that teachers used a whole-class lecture style of teaching, mostly for explaining concepts. Although the teachers attempted to demonstrate mathematical concepts dynamically, either they tended to use the technology statically or they avoided using it. The teachers mostly used the IWB as a non-digital whiteboard.

1 Introduction

For technology to introduce innovation into the teaching and learning of mathematics, we as researchers need to understand the everyday practice of teachers and to suggest new means of technological implementation (Clark-Wilson, Sinclair, & Robutti, 2013). In recent years many studies have examined mathematics teaching in technological environments that are accessible to both the teacher and the students. Some of these studies focused on characterizing teachers' knowledge needed for integrating

M. Tabach (✉) · G. Slutzky
Tel Aviv University, Tel Aviv-Yafo, Israel
e-mail: tabachm@post.tau.ac.il; tabach.family@gmail.com

G. Slutzky
e-mail: galit.slutzky@gmail.com

E. Faggiano et al. (eds.), *Innovation and Technology Enhancing Mathematics Education*, Mathematics Education in the Digital Era 9,
DOI 10.1007/978-3-319-61488-5_10

technology into their everyday routine (e.g., Mishra & Koehler, 2006). Other researchers focused on the ongoing documentation work of the teachers, done mainly outside the classroom, while re-sourcing digital and other sources into his everyday practice (e.g., Chap. Innovations Through Institutionalized Infrastructures: The Case of Dimitris, His Students and Constructionist Mathematics; Pepin, Gueudet, and Trouche, (eds.), 2013). Others identified the unique role played by the teacher in the mathematics classroom where students have technological means at their disposal. Specifically, the instrumental orchestration framework proposed by Trouche (2004) and developed by Drijvers and colleagues (Drijvers, Doorman, Boon, Reed, & Gravemeijer, 2010; Drijvers, Tacoma, Besamusca, Doorman, & Boon, 2013) is relevant in cases where both teacher and students have access to technology. Yet in many high schools in Israel, and likely elsewhere in the world as well, classrooms are equipped only with a computer for the teacher and a data projector. The teachers do not consider students accessibility to smartphones as a possible technological resource to be used in class (see, for example, Chap. Innovative Uses of Digital Technology in Undergraduate Mathematics; Trouche & Drijvers, 2010). That is, the technology is available only to the teacher. Hence, in order to analyze teacher practice in cases where students do not have access to technology in class, we need to modify the existing framework or to develop an innovative one.

The current study monitored several second-year mathematics teachers with strong technological backgrounds. We chose these specific teachers intentionally because the claim that teachers lack technological knowledge does not apply in their case. Hence it will be interesting and innovative to trace their use of technology in class. The classrooms were equipped with only one computer for the teacher. In some cases the classroom was also equipped with an interactive white board. The study sought to develop a framework that would allow us to characterize teachers' actions in such settings and to examine the applicability of the framework by applying it to several case studies. The study also investigated teacher knowledge to examine whether there is any connection between this knowledge and teacher actions. The use of several case studies allowed us to highlight similarities and differences between cases.

2 Theoretical Framework

2.1 *Integrating Technology into Mathematics Lessons*

Koehler and Mishra (2008) define technology in the school context as "the sum of the tools, techniques, and collective knowledge applicable to education" (p. 5). This definition does not distinguish between what may be considered "old" technology, such as blackboards or overhead projectors, and "new" technology, such as Internet applets. Some technologies, such as chalk for example, are more easily classified as "old." However, with the rapid advances in digital technology, the classification of technology as "new" is bound to time and place. Furthermore, in some classrooms "old" and "new" technologies are often used at the same time.

Every technology has specific affordances and constraints (Gibson, 1979). In the context of educational technology, these affordances and constraints refer to all the properties of a system that allow certain actions to be performed and encourage specific types of learner behaviors (Norman, 1988). A distinction can be made between affordances and constraints that are part of the technological tool itself and those that are imposed by the task or the user. A teacher may refrain from giving her lecture notes to her students due to her personal "functional fixedness" (German & Barrett, 2005), even though as a tool these notes may provide her students another learning opportunity. Sometimes the integration of technology is dependent upon overcoming functional fixedness. Specifically, in the classroom the teacher is often the one who needs to overcome her practice fixedness in order to harness the technology's affordances for the benefit of student learning. Tabach (2011) reported on a case of one skillful teacher who worked in a particular technological environment for some time before she was able to overcome her own fixedness regarding the use of technology.

Teacher practice is complex and stable (Robert & Rogalski, 2005), to some extent due to teachers' functional fixedness. Integrating technology amplifies this complexity and challenges the stability of teacher practice because new techniques need to be developed for integrating technological tools (Lagrange & Monaghan, 2009). Such new techniques are likely to be related to already existing ones as well as to teachers' perceptions about mathematics education (Pierce & Ball, 2009).

2.2 *Technological Pedagogical Content Knowledge (TPACK)*

The notion of TPACK emerged in an attempt to propose a typology of teacher knowledge for technology integration based on Shulman's (1986) construct of pedagogical content knowledge (PCK). Shulman (1986) explained his ideas about PCK as follows:

> Pedagogical content knowledge is of specific interest because it identifies the distinctive bodies of knowledge for teaching. It represents the blending of content and pedagogy into an understanding of how particular topics, problems or issues are organized, represented, and adapted to the diverse interests and abilities of learners, and presented for instruction (p. 8).

The construct of PCK was criticized by Ball and Bass (2000), who claimed that "it sometimes falls short in the dynamic interplay of content in teacher's real-time problem solving (p. 88)." Yet this notion has shaped views of teacher knowledge for the last three decades (Graeber & Tirosh, 2008). Shulman's ideas evolved in an environment in which the technological tools used in class, among them textbooks, overhead projectors and the like, were considered commonplace and as such transparent (Bruce & Hogan, 1998). The introduction of digital computers and software into school systems and the increasing availability of computer communications technology brought the issue of technology to the fore. One may argue that Shulman's construct of Curricular Knowledge (CK) includes technology. Shulman (1986) defines CK as follows:

> The curriculum is represented by the full range of programs designed for the teaching of particular subjects and topics at a given level, the variety of instructional materials available in relation to those programs, and the set of characteristics that serve as both the indications and contraindications for the use of particular curriculum or program materials in particular circumstances. … [W]e ought to expect that the mature teacher possesses such understanding about the curricular alternatives available for instruction (p. 10).

Yet as Mioduser (1998) explained, "[t]echnology is generally presented as a discipline in its own right, as a body of knowledge with its unique history and development, philosophy, contents, and methodology" (p. 169). If we accept this view, we must acknowledge that Shulman's typology has a missing component that needs to be explicitly addressed as a component in teachers' knowledge.

Shulman's (1986) seminal work suggested that content (C) and pedagogy (P) are two bodies of knowledge that must overlap, creating a third distinct body of knowledge (PCK). Similarly, Mishra and Koehler (2006) suggested that technological knowledge (T) also needs to overlap knowledge regarding the content to be learned and the pedagogy (Fig. 1). By accepting this complex view of teacher

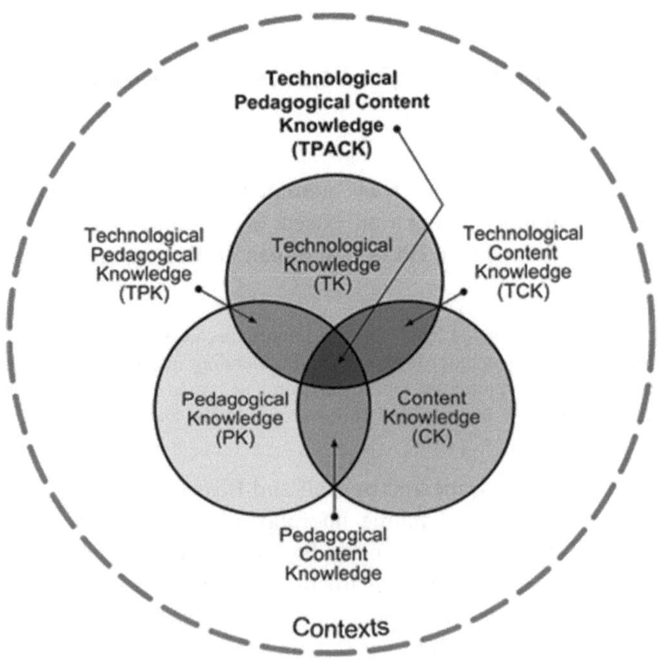

Fig. 1 Diagram of the combined knowledge areas, copied with permission from http://tpack.org/

knowledge, we must take into consideration four additional distinct yet connected bodies of knowledge.

Mishra and Koehler (2006) proposed the following descriptions for each of these four new bodies of knowledge. Technological knowledge (TK) entails the teacher's knowledge about digital computation devices and software. This includes knowledge regarding how to connect and install as well as how to use hardware and software. An important characteristic of TK is the ability to learn about new technologies. To illustrate, one may consider the ability to adapt from one version of software to its successive version. Technological pedagogical knowledge (TPK) refers to knowledge relating to how the use of technology may influence teacher practice and student learning. It includes familiarity with existing tools for a specific pedagogical purpose and knowing how to choose an appropriate teaching strategy according to the learning goals. Technological content knowledge (TCK) involves knowing how technology and content mutually influence each other. For example, dynamic geometry software allows students to construct and "drag" a construction so they can inspect many examples of a constructed object. In a sense, the tool provides students the opportunity to experiment so that the dynamic geometry software serves as a laboratory.

Technological pedagogical content knowledge (TPCK) is the knowledge at the core where pedagogical knowledge, content knowledge and technological knowledge overlap. Mishra and Koehler (2006, p. 1029) define TPCK according to the following five characteristics: [1] an understanding of the representation of concepts using technology; [2] pedagogical techniques that use technologies in constructive ways to teach content; [3] knowledge of what makes concepts difficult or easy to learn and how technology can help redress some of the problems that students face; [4] knowledge of students' prior knowledge and theories of epistemology; and [5] knowledge of how technologies can be used to build on existing knowledge and to develop new epistemologies or strengthen old ones.

Figure 1 illustrates the view that these combined knowledge areas are bounded within contexts. In other words, integrated knowledge is strongly related to subject, though it needs to be further elaborated for each subject area.

Many researchers have used TPCK in their studies both within mathematics education and in other educational domains. Voogt, Fisser, Pareja Roblin, Tondeur & van Braak (2013) reviewed and analyzed the ways in which different researchers interpreted the TPCK framework. They pointed to three different ways of understanding the TPCK concepts: T(PCK) as an extension of PCK by integrating Technological Knowledge; TPCK as a unique and distinct body of knowledge; and TP(A)CK as a knowledge domain that emerges from the integration of Technological Knowledge, Pedagogical Knowledge and Content Knowledge. While the first two conceptualizations view TPCK as a knowledge domain on its own, TP(A)CK represents an integrative view and emphasizes the relationship between the three knowledge domains and their intersections. This is the view of TPACK adopted in the current study.

2.3 Ways of Analyzing Teacher Practice in Technological Environments

Vérillon and Rabardel (1995) defined instrumental genesis as the process by which individuals create and change their perceptions of a tool while performing different tasks. Instrumental genesis is considered a bidirectional process in which both tool and user change. Trouche (2004) referred to these two bidirectional aspects as instrumentalization and instrumentation. Changes in the tool, or instrumentation, usually entail changes in software features. For example, user-generated changes to cellular devices reflect users' needs and their orientation towards the tool. Changes in the user, known as instrumentalization, affect the mental schema developed by users with respect to achieving a mathematical goal with the aid of a specific technological tool.

Whole-class discussions orchestrated by the teacher (Trouche, 2004) can serve as an appropriate forum for discussing and sharing students' own instrumental geneses for the purpose of further enhancement. Trouche "introduced the term *instrumental orchestration* to point out the necessity of *external steering* of students' instrumental genesis" (2004, p. 296, emphasis in the original). Rivera put it differently: "Instrumental orchestration encompasses institutional strategies that assist students in developing instrumental actions. The strategies are aimed at learners who are driven by the goals in an activity" (Rivera, 2007, p. 295). Instrumental orchestration also has a socio-cultural aspect (Laborde, 2003; Lagrange et al., 2003), since the technological medium serves as a boundary object between teacher and students, where "mutual negotiation and meaning-construction is the norm for both sides" (Hoyles et al., 2004, p. 321).

> Instrumented orchestration is defined by four components: a set of individuals; a set of objectives (related to the achievement of a type of task or the arrangement of a work-environment); a didactic configuration (that is to say a general structure for the plan of action); a set of exploitations of this configuration (Guin, Ruthven, & Trouche, 2005, p. 208).

The framework of instrumental orchestration does not suggest specific orchestration types. Nevertheless, several orchestration types have been identified based on empirical data from various studies (Drijvers et al., 2010, 2013; Tabach, 2011, 2013; Trouche, 2004).

The framework of instrumental orchestration was suggested by Trouche and developed by others in an attempt to better understand teacher practice in mathematics classrooms with digital tools accessible to students. Nevertheless, in many schools today students are not equipped with digital technology for use in the context of mathematics lessons. If only the classroom teacher is equipped with a computer and data projector and the students do not have access to digital technology, applying the instrumental orchestration framework to teacher practice is not adequate. A different framework is needed to be able to describe what happens.

3 Methods

3.1 Research Question

Four case studies were examined in light of four questions. The first two questions refer to each case separately, while the last two compare the four cases to highlight similarities and differences. Each case focused on one high school mathematics teacher working in a technological environment comprising one computer and a data projector.

1. What teaching practices can be identified in class?
2. What are the teachers' TPACK knowledge components?
3. What similarities and differences can be identified in the practices of the teachers?
4. What similarities and differences can be identified in teacher knowledge?

3.2 Participants and Research Procedure

Four high school teachers—Meir,[1] Gilad, Eli and Ron—took part in the study. These four teachers had several years of experience working in high-tech industry. Each of them decided to change careers and become high school mathematics teachers. They all studied in the same year-long program designed for people who hold bachelor's degrees in mathematics and wish to become high school mathematics teachers. Note that we had no involvement in their professional training program. We felt that the similar backgrounds of the four teachers would provide us with a good starting point for comparison. At the time of the research, all the teachers had already been teaching high school mathematics in different schools for about a year. The classrooms in all four schools were equipped with a computer and a data projector for teacher use. Two of the schools also had interactive white boards in the classrooms.

A preliminary discussion took place with each teacher to explain the research goals and obtain consent to participate in the study. All four teachers used technology in their teaching on a regular basis. We used individual semi-structured interviews to interview each of the teachers in a quiet corner for about an hour. The goal of these interviews was to gain some insights into each teacher's knowledge. Each teacher was observed during three to four lessons in one of his regular classes, based on his choice of a convenient time. The interviews and observations were recorded and transcribed verbatim.

[1]All names are pseudonyms.

3.3 Data Analysis

Based on observations in classrooms where technology was available only to the teacher, we propose a framework to examine the practices of these teachers. The framework includes three lenses: (1) the teachers' goals; (2) the technological resources used; and (3) the way the resources were used. The teachers' goals in the first lens were categorized as follows: proving a statement; explaining a concept, technique or how to solve a problem; monitoring and guiding in which students work individually while the teacher circulates and helps as needed; and management, in which all kinds of general actions take place, such as school announcements. Note that these goals were identified based on observing the teachers acting in the classes. Table 1 outlines the operational definitions for each category.

The second lens, technical resources, was not always applicable, for example in cases when the data projector was not used. When the data projector was in use, the teacher could choose to work with mathematical software or general software. Table 2 outlines the operationalization of technical resources.

Table 1 Operationalization of teachers' goals

Goal		Means of identification
Proving (Prov)		The teacher proves a mathematical statement
Explaining	Concept (ExpC)	The teacher explains a mathematical idea, concept or claim
	Technique (ExpT)	The teacher explains a general technique or algorithm for solving a particular type of problems
	How to solve (H2S)	The teacher explains how to solve a particular problem or rehearses a particular topic in the whole-class forum
Monitoring & Guiding (M&G)		The teacher circulates among the students who work as individuals or in pairs. The teacher assists in private conversations as the need arises
Management (Mng)		The teacher does administrative work, such as informing students about events, giving information and the like

Table 2 Operationalization of technical resources

Software type	Means of identification
Mathematical software (MS)	Software that allows the user to manipulate mathematical objects, such as symbolic expressions, graphs, geometrical figures, calculations and others. Examples include Excel, GeoGebra
General software (GS)	Software that does not allow the user to manipulate mathematical objects. Sometimes the software may display a mathematical object. Examples include Word, Acrobat Reader, PowerPoint
NA	The use of technical resources was not applicable (NA) if the data projector was turned off

Table 3 Operationalization of way resources were used

Ways of use	Means of identification	Distinction
Static presentation (Stat)	Using software to display a document, picture or their integration into one page (or more, e.g., PPT), similar to the use of an overhead projector	Not possible to manipulate or interact
Presentation includes multi-media (MM)	Using manipulatives or demonstrations including multimedia, such as animations, movies, simulations	
Applet	Use of demonstrative digital tools that can be manipulated or interacted with by touch, keyboard or mouse	Possible to manipulate or interact
Learning management or context system (Org Teach)	Projecting software onto board that can be used to organize or edit content. Students watch and may influence the end product. Examples: Word or PowerPoint presentation, or editing abilities of an interactive white board	
Regular board uses	The teacher can write on the projected display (WO), write beside the projected display (WB) or not write on the display (NW)	

The third lens refers to the way the resources were used and includes four categories. The first distinction refers to the teacher's ability to manipulate or interact with the display, including static presentation or presentation that includes a multimedia component. In other cases, the teacher manipulates the presentation or interacts with it, including the use of applets or learning management organizing systems (see Table 3).

We also operationalized the component of the TPACK framework to analyze teachers' knowledge as it emerged from the classroom observations and was expressed in the individual interviews with the teachers. Table 4 provides some examples for categorizing teachers' utterances during the interview into knowledge types. Further, we categorized the questions asked during each interview according to the knowledge types they were expected to elicit. We then compared the knowledge types aimed at in each question with the knowledge types we categorized in the teachers' answers. Mismatches between the knowledge types aimed at and those answered allowed us to determine the level of the particular knowledge type. For example, if the teachers' answers to questions aiming to elicit pedagogical knowledge included almost no utterances we could classify this as pedagogical knowledge of a low level. In contrast, if the utterances in the answers were classified as technological knowledge regardless of the knowledge type the questions aimed to elicit, we determined this to be a very high level knowledge type.

Table 4 Operationalization for identifying teachers' knowledge

Knowledge types	Means of identification	Examples
TK	Teacher operates software or hardware; can handle malfunctions	Where to save files How to operate the software
PK	Teacher refers to general pedagogical issues, with no reference to mathematics	Keeping order in class Reading names
CK	Teacher refers to mathematical facts, procedures, process, proofs…	What is a derivative? How to solve quadratic equation
PCK	Teacher refers to ways of teaching mathematics, considerations of pedagogical choices regarding sequencing learned topics and more	What may affect students' learning of a particular topic? What is considered easy or hard for students? Students' difficulties or misconceptions Students' prior knowledge that may influence current learning Best examples, analogies, and the like
TPK	Teacher refers to operating technology for integration in teaching with no reference to mathematics	How to use the interactive white board for teaching Pros and cons of using technology while teaching
TCK	Teacher refers to operating technology for presenting mathematical content	Explaining about an applet for learning math
TPACK	Teacher refers to using the software for teaching mathematics	How to choose a digital representation appropriate to the learned topic How to choose a pedagogical technique that will make use of digital tools to study a specific math topic How can specific problematic issues for students be overcome with the help of technology

4 Findings

First we present the findings from each teacher's practice separately. After that we analyze the knowledge regarding the four teachers together.

4.1 Meir's Teaching Method

Meir's classroom was equipped with an interactive white board and a regular white board, but he used only the interactive white board both for data projecting and for writing. He used Geogebra files he created at home or drawings he created on the spot with Geogebra in response to students' needs. When he wrote on the

Table 5 Distribution of Meir's teaching during the observed lessons

Teaching goals		Technological resources		Use of resources	
	%		%		%
Explain concept	56	Mathematical software	42	Applet	42
		General software	14	Org Teach	14
Explain technique	2	Mathematical software	2	Org Teach	2
Explain how to solve	24	General software	24	Org Teach	24
Management	18	General software	12	Static	12
		NA	6		

*All percentages are rounded to whole numbers

interactive white board he used it as a regular white board, making no use of its software abilities. Table 5 summarizes the analysis of Meir's observed lessons according to teacher goals. During almost half the lesson time (44%) Meir used mathematical software, and for the vast majority of the lesson time (88%) he used the technology dynamically while explaining mathematical concepts, techniques and ways to solve exercises.

4.2 Gilad's Teaching Method

Gilad's classroom was also equipped with two boards—an interactive white board and a regular one. Like Meir, he also only used the interactive white board both for data projecting and for writing. Like Meir, he used Geogebra files he created at home or drawings he created on the spot in response to students' needs. Unlike Meir, when Gilad wrote on the interactive white board he used its software capabilities (colored changes, moving, scaling, drawing lines and the like). Table 6 summarizes the analysis of Gilad's observed lessons according to teacher goals. During about a fifth of the lesson time (19%) Gilad used mathematical software, yet for the vast majority of the lesson time (75%) he used the technology dynamically. He spent less time on management issues compared to Meir, and he also spent a short time proving a statement, which did not occur in Meir's observed lessons.

4.3 Eli's Teaching Method

Eli's class was equipped only with a regular white board on which the data were projected. Eli also used Geogebra files that he created at home or drawings he created on the spot with Geogebra in response to students' needs. While the data were projected on the white board, he used an erasable marker pen to write on and next to the display. Table 7 summarizes the analysis of Eli's observed lessons

Table 6 Distribution of Gilad's teaching during the observed lessons

Teaching goals		Technological resources		Use of resources	
	%		%		%
Prove a statement	6	General software	6	Org Teach	6
Explain concept	58	Mathematical software	18	Stat	8
				Applet	10
		General software	40	Stat	13
				Org Teach	27
Explain technique	11	Mathematical software	1	Applet	1
		General software	10	Org Teach	10
Explain how to solve	13	General software	13	Org Teach	13
Management	12	General software	5	Stat	4
				Org Teach	1
		NA	7		

*All percentages are rounded to whole numbers

Table 7 Distribution of Eli's teaching during the observed lessons

Teaching goals		Technological resources		Use of resources		Teacher writes on board	
	%		%		%		%
Prove a statement	9	Mathematical software	9	Stat	2	WO & WB	2
				Applet	7	WO & WB	7
Explain concept	46	Mathematical software	44	Stat	25	WO	7
						WB	11
						WO& WB	7
				Applet	19	WO	12
						WO &WB	6
						NW	1
		NA	2				
Explain how to solve	36	NA	36				
Monitor & guide	6	Mathematical software	6	Stat	6	NW	6
Management	3	NA	3				

*All percentages are rounded to whole numbers; *WO* Writes on the projected display; *WB* Writes beside the projected display; *NW* Does not write on the display according to teacher goals. Eli used mathematical software during 59% of the lesson time, a bit more than half of that time in a static mode while writing on and next to the displayed data. He spent significantly less time on management issues. Like Gilad, he also spent a short time proving a mathematical statement. Finally, he engaged his students as individuals or in pairs for a short period during class work while he circulated among them.

4.4 *Ron's Teaching Method*

Ron's classroom was organized similar to Eli's and was equipped with a regular white board on which the data were projected. Ron used only a digital version of the student textbook, which he projected onto the white board. When the data were projected onto the board, he used an erasable marker pen to write on and next to the display. Table 8 summarizes the analysis of Ron's observed lessons according to teacher goals. Ron used general software for the vast majority of the lesson time, 87% in a static mode only. During almost all that time he wrote on and next to the displayed data. The amount of time he used for management issues was similar to that used by Meir and Gilad.

Table 8 Distribution of Ron's teaching during the observed lessons

Teaching goals		Technological resources		Use of resources		The teacher writes on board	
	%		%		%		%
Explain concept	9	NA	1				
		General software	8	Stat	8	NW	4
						WB	3
						WO& WB	1
Explain technique	2	General software	2	Stat	2	NW	2
Explain how to solve	73	General software	73	Stat	73	NW	1
						WO	21
						WB	37
						WO & WB	14
Management	16	NA	12				
		General software	4	Stat	4	NW	4

*All percentages are rounded to whole numbers; *WO* Writes on the projected display; *WB* Writes beside the projected display; *NW* Does not write on the display

Table 9 Teachers' knowledge

	TK	PK	CK	PCK	TPK	TCK	TPACK
Meir	*Strong*	Low	*Strong*	Low	*Strong*	*Strong*	*Strong*
Gilad	*Strong*	Medium	*Strong*	Medium	*Strong*	*Very strong*	*Very strong*
Eli	*Strong*	Medium	*Strong*	*Strong*	*Strong*	*Strong*	*Strong*
Ron	*Strong*	Low	*Strong*	*Strong*	Low	Low	Medium

4.5 Findings Regarding Knowledge for All Four Teachers

As mentioned above, the teachers' knowledge was analyzed based on the interviews. Table 9 shows the overall results regarding all four teachers. The teachers participating in the study were not typical. On the one hand, they had a high level of technological and mathematical knowledge due to their previous work experience in high-tech industry. We can see that for the most part, their TPK, TCK and TPACK are on a high level. We attribute this high level to their strong technological knowledge component. On the other hand, as practicing teachers in their first or second year of teaching, their pedagogical knowledge is more limited. This is also to some extent reflected in their PCK. This last point is of interest, as all four teachers are graduates of the same year-long training program. It seems that both their professional training and the enculturation process at the schools where they teach resulted in different levels of knowledge for the various knowledge types.

5 Discussion

The current study used a multiple-case approach to examine teacher practice in a specific technological environment in which only one computer and one data projector were available in the classroom. Each of the four teachers was interviewed and observed for three to four lessons in one class. The study sought to identify teachers' knowledge and practice and possibly to highlight the links between the two.

Examining teachers' knowledge via the TPACK framework indicated on the one hand that strong technological knowledge and strong content knowledge do not necessarily indicate strong technological content knowledge. Nevertheless, strong technological knowledge may compensate for low pedagogical knowledge in the technological pedagogical knowledge component. Therefore, we conceptualize each of the seven knowledge components pointed out by Mishra and Kohler (2006) as a body of knowledge in its own right, with multiple connections among these knowledge components. Voogt et al. (2013) referred to this view of multiple connections and interdependency as TP(A)CK. Our study provides empirical evidence for this theoretical point of view.

To examine teacher practice, we developed an innovative three-lens framework that we used concurrently to identify similarities and differences in the practices of the four teachers described. The framework allowed us to gain some insights into teacher practice and hence proved to be a useful research tool. In the following we discuss each lens and its possible connections to teacher knowledge.

In terms of the teachers' learning goals, we can see from Table 10 that all four teachers used explanations—of concepts, techniques or ways to solve—for about 80% of their teaching time. In other words, the lessons were mainly teacher-centered. This is in accordance with the findings of Zevenberg and Lerman (2008), who studied teacher practice in an interactive white board environment. These researchers claimed that the potential of the interactive white board to foster new pedagogies remains unfulfilled and that teachers prefer the whole-class setting. Further, they claimed that sometimes teachers are engaged with the technical aspects of how to use the interactive white board, which hinders meaningful discussions. In the current study, we showed that this was also the case for the two teachers who did not have an interactive white board but only a data projector and a regular white board. Their dominant teaching style was the same. We cannot claim that these four teachers were accustomed to teaching in a whole-class lecture style prior to the introduction of technology to their class as they began as teachers who used technology. We can attribute this pedagogical choice to "functional fixedness" (German & Barrett, 2005), which evolved during all the years they were studying and observing their own teachers, as well as to their limited pedagogical knowledge.

When considering the findings shown in Tables 9 and 10, we can establish a link between the teachers' ways of teaching and their pedagogical knowledge. The four teachers had low-medium pedagogical knowledge, and they all used traditional teacher-centered pedagogy while teaching. Only for a short period of time in Eli's lessons did the students worked individually or in pairs while he circulated among them.

All the teachers used technology, either statically or to present a document (Table 11). In addition, some of the teachers used applets. The two teachers who had an interactive white board at their disposal used the technology to organize learning content. We did not see the teachers using multimedia in any of the

Table 10 Distribution of teaching goals (in percentages) for each teacher

Teaching goals		Meir	Gilad	Eli	Ron
Proving (Prov)		–	6	9	–
Explain	Concept (ExpC)	*56*	*58*	46	9
	Technique (ExpT)	*2*	*11*	–	2
	How to solve (H2S)	*24*	*13*	36	73
Monitor & Guide (M&G)		–	–	6	–
Management (Mng)		*18*	*12*	3	16

Shaded cells for teaching in an IWB class

Shaded cells for teaching in an IWB class

Table 11 Distribution of teachers' use of technology (in percentage) for each teacher

Methods of use	Meir	Gilad	Eli	Ron
Static presentation (Stat)	*12*	*25*	33	87
Presentation includes multi-media (MM)	–	–	–	–
Applet	*42*	*11*	26	–
Learning management or context system (Org Teach)	*40*	*57*	–	–
Regular board uses	*6*	*7*	41	13

Shaded cells for teaching in an IWB class

Shaded cells for teaching in an IWB class

observed lessons, though the teachers claimed to have used multimedia. Kaput (1992) highlighted the importance of visual and dynamic representations for mathematical concepts, claiming that such representations help learners build a mental representation of mathematical concepts. Most of the teachers in the current study tried to use Geogebra as a means of incorporating visual and dynamic representations, but in fact they spent much of the time in a static mode. Little use of dynamism was also reported by Ruthven (2009), who studied an unusually expressive teacher while teaching high school students in a dynamic geometry environment. This is also in line with the findings of Hofer and Harris (2010), who claim that integrating technology into teacher practice is a complex process.

In addition, Table 11 shows that the teachers who taught using an interactive white board spent less time not using technology (the not applicable category). In contrast, the two teachers without an interactive white board did not use software to edit content, as in Word editing. Eli's NA time use is similar to the time used by Meir and Gilad to organize teaching time. In fact, Meir and Gilad used the interactive white board to write the mathematics content, while Eli did the same on the regular board. This is in line with the findings of Drijvers et al. (2013) that the board-instruction category, in which teachers used the board as if no technology is available, is common. These researchers claim that this way of overlooking technology allowed the teachers to remain close to their old ways of teachings.

The one exception was Ron. Even though he did not use an interactive white board, he used technology in almost all the observed lesson time. He is also exceptional in his static use of the same technology (digital book) almost all the time.

The three teachers who exhibited strong TCK—Meir, Gilad and Eli—also used mathematical software while teaching (Table 12). Eli was the only teacher who did not use general software. Meir and Gilad used the interactive white board software to organize mathematical content, while Ron used the digital book display.

We conclude with a practical observation relevant for teacher training and with three innovative research observation. The first innovative feature of the current study was the technological environment. As students had no access to technology in the classroom, the well-developed frameworks were not applicable, and we had to find a way to focus our observations on teacher practice. More examination is needed to evaluate the benefits and pitfalls of the suggested framework.

Table 12 Distribution of technical resources used (in percentage) by each teacher

Software type	Meir	Gilad	Eli	Ron
Mathematical software (MS)	*44*	*19*	58	–
General software (GS)	*50*	*74*	–	87
NA	*6*	*7*	41	13

Shaded cells for teaching in an IWB class

Shaded cells for teaching in an IWB class

The second innovation relates to the unique group of participants in the study. The unique teacher population that was the focus of the current study reflects a growing trend of people who have chosen a second career as teachers after a period of working in high-tech industry. This population has special characteristics in terms of their strong background of technological and mathematical knowledge. Their weakness lies in their pedagogical knowledge. Hence, professional development programs aimed at this particular group should focus on pedagogical knowledge, pedagogical content knowledge and the integration of pedagogy, content and technology—TPACK.

Finally, other studies also noted a gap between the potential of integrating technology into school mathematics and the actual use teachers make of this technology in class. The literature suggests two common explanations: (1) teachers need to change their practice from working in a traditional environment to a digital environment, and such a change is highly complex (Robert & Rogalski, 2005; Zevenbergen & Lerman, 2008); and (2) teachers' technological knowledge is not sufficiently developed, and hence they do not use the full potential of the digital tools (Lagrange & Monaghan, 2009). The innovative power of the current study is that in the four analyzed cases, both explanations fail. The teachers were in their second year as teachers and hence did not have well established practices to change. Moreover, the teachers had strong technological knowledge. Perhaps we as a research community still need to find better ways to understand teacher practice and how to help teachers use the full potential of digital environments.

References

Ball, D. L., & Bass, H. (2000). Interweaving content and pedagogy in teaching and learning to teach: Knowing and using mathematics. In J. Boaler (Ed.), *Multiple perspectives on the teaching and learning of mathematics* (pp. 83–104). Westport, CT: Ablex.

Bruce, B. C., & Hogan, M. C. (1998). The disappearance of technology: Toward an ecological model of literacy. In D. Reinking, M. McKenna, L. Labbo, & R. Kieffer (Eds.), *Handbook of literacy and technology: Transforming in a post-typographic word* (pp. 269–281). Hillsdale, NJ: Erlbaum.

Clark-Wilson, A., Sinclair, N., & Robutti, O. (Eds.). (2013). *The mathematics teacher in the digital era*. Dordrecht: Springer.

Drijvers, P., Doorman, M., Boon, P., Reed, H., & Gravemeijer, K. (2010). The teacher and the tool: instrumental orchestrations in the technology-rich mathematics classroom. *Educational Studies in Mathematics, 75*(2), 213–234.

Drijvers, P., Tacoma, S., Besamusca, A., Doorman, M., & Boon, P. (2013). Digital resources inviting changes in mid-adopting teachers' practices and orchestrations. *ZDM Mathematics Education, 45*(7), 987–1001.

German, T., & Barrett, C. (2005). Functional fixedness in a technologically sparse culture. *Psychological Science, 16*(1), 1–5.

Gibson, J. J. (1979). *The ecological approach to visual perception*. Boston: Houghton Mifflin.

Graeber, A., & Tirosh, D. (2008). Pedagogical content knowledge: A useful or an elusive notion? In P. Sullivan (Ed.), *Knowledge and beliefs in mathematics teaching and teaching development* (pp. 117–132). Amsterdam, The Netherlands: Sense.

Guin, D., Ruthven, K., & Trouche, L. (Eds.). (2005). *The didactical challenge of symbolic calculators: Turning a computational device into a mathematical instrument*. New York: Springer.

Hofer, M., & Harris, J. (2010). Differentiating TPACK development: Using learning activity types with inservice and preservice teachers. In C. D. Maddux, D. Gibson, & B. Dodge (Eds.), *Research highlights in technology and teacher education 2010* (pp. 295–302). Chesapeake, VA: Society for Information Technology and Teacher Education (SITE).

Hoyles, C., Noss, R., & Kent, P. (2004). On the integration of digital technologies into mathematics classrooms. *International Journal of Computers for Mathematical Learning, 9*(3), 309–326.

Kaput, J. (1992). Technology and mathematics education. In D. A. Grouws (Ed.), *Handbook of research on mathematics teaching and learning* (pp. 515–556). Reston, VA: National Council of Teachers of Mathematics.

Koehler, M. J., & Mishra, P. (2008). Introducing TPCK. In AACTE Committee on Innovation and Technology (Eds.), *Handbook of technological pedagogical content knowledge (TPCK) for educators,* (pp. 3–30). NY, USA: Routledge.

Laborde C. (2003). Technology used as a tool for mediating knowledge in the teaching of mathematics: The case of Cabri-geometry. In W-C. Yang, S.C. Chu, T. de Alwis, Ming Gong Lee (Eds.), *Proceedings of the 8th Asian Technology Conference in Mathematics* (Vol. 1, pp. 23–38). Hsinchu, Taiwan ROC: Chung Hua University.

Lagrange, J. B., Artigue, M., Laborde, C., & Trouche, L. (2003). Technology and mathematics education: Multidimensional overview of recent research and innovation. In A. J. Bishop, M. A. Clements, C. Keitel, J. Kilpatrick, & F. K. S. Leung (Eds.), *Second international handbook of mathematics education* (Vol. 1, pp. 237–270). Dordrecht: Kluwer Academic Publishers.

Lagrange, J. B., & Monaghan, J. (2009). On the adoption of a model to interpret teachers' use of technology in mathematics lessens. In V. Durand- Guerrier, S. Soury-Lavergne, & F. Arzarello (Eds.), *Proceedings of the Sixth Congress of the European Society for Research in Mathematics Education* (pp. 1605–1614). Lyon: INRP. Retrieved on April, 4th, 2011, from http://www.inrp.fr/editions/editions-electroniques/cerme6/working-group-9

Mioduser, D. (1998). Framework for the study of the cognitive nature and architecture of technological problem solving. *Journal of Technology Education and Design, 8*(2), 167–184.

Mishra, P., & Koehler, M. J. (2006). Technological pedagogical content knowledge: A framework for teacher knowledge. *Teachers College Record, 108*(6), 1017–1054.

Norman, D. A. (1988). *The psychology of everyday things*. New York: Basic Books.

Pepin, B., Gueudet, G., & Trouche, L. (eds.) (2013). Re-sourcing teacher work and interaction: new perspectives on resource design, use and teacher collaboration, special issue of *ZDM. The International Journal on Mathematics Education, 45*(7), 929–944.

Pierce, R., & Ball, L. (2009). Perceptions that may affect teachers' intention to use technology in secondary mathematics classes. *Educational Studies in Mathematics, 71*(3), 299–317).

Rivera, F. (2007). Accounting for students' schemes in the development of a graphical process for solving polynomial inequalities in instrumented activity. *Educational Studies in Mathematics, 65,* 281–307.

Robert, A., & Rogalski, J. (2005). A cross-analysis of the mathematics teacher's activity. An example in a French 10th-grade class. *Educational Studies in Mathematics, 59*(1–3), 269–298.
Ruthven, K. (2009). Towards a naturalistic conceptualisation of technology integration in classroom practice: The example of school mathematics. *Education & Didactique, 3*(1), 131–149.
Shulman, L. S. (1986). Those who understand: Knowledge growth in teaching. *Educational Researcher, 15*(2), 4–14.
Tabach, M. (2011). A mathematics teacher's practice in a technological environment: A case study analysis using two complementary theories. *Technology, Knowledge and Learning, 16*(3), 247–265.
Tabach, M. (2013). Applying instrumental orchestration framework to different educational institutions. *CERME8- Eight Conference of European Research in Mathematics Education.* Retrieved February, 14, 2013, from http://www.cerme8.metu.edu.tr/wgpapers/WG15/WG15_Tabach.pdf
Trouche, L. (2004). Managing the complexity of the human/machine interaction in computerized learning environments: guiding students' command process through instrumental orchestrations. *International Journal of Computers for Mathematical Learning, 9*(3), 281–307.
Trouche, L., & Drijvers, P. (2010). Handheld technology for mathematics education: flashback into the future. *ZDM, The International Journal on Mathematics Education, 42*(7), 667–681. doi:10.1007/s11858-010-0269-2
Vérillon, P., & Rabardel, P. (1995). Cognition and artifact: A contribution to the study of thought in relation to instrumented activity. *European Journal of Psychology in Education, 9*(3), 1–33.
Voogt, J., Fisser, P., Pareja Roblin, N., Tondeur, J., & van Braak, J. (2013). Technological pedagogical content knowledge—A review of the literature. *Journal of Computer Assisted learning, 29*(2), 109–112.
Zevenbergen, R., & Lerman, S. (2008). Learning environments using interactive whiteboards: New learning spaces or reproduction of old technologies? *Mathematics Education Research Journal, 20*(1), 108–126.

Part V
Closing Scenery

Digital Mazes and Spatial Reasoning: Using Colour and Movement to Explore the 4th Dimension

Elizabeth de Freitas

Abstract This chapter focuses on innovative developments of four-dimensional digital mazes, examining how these mazes tap into the ideas of mathematician and fiction writer Charles Hinton (1853–1907) who wrote extensively on perception of a 4th *geometric* dimension. Hinton treats mathematical objects as physical and material movements, and draws on non-Euclidean geometry to argue for a virtual dimension to matter. I discuss recent attempts to build digital mazes that develop spatial sense in four dimensions, and show how these are directly linked to Hinton's ideas. I focus on how colour and movement in digital environments are used to develop a distinctive kind of spatial sense. This chapter sheds light on innovative uses of digital software for developing student spatial sense. My aim is to explicate the new materialism of Charles Hinton, contribute to discussions about the nature of spatial sense and spatial reasoning, and to point to possible directions for future research on inventive approaches to geometry.

1 Introduction

Charles Howard Hinton (1853–1907) was a British mathematician and author who published various monographs on the mathematics of higher dimensions. He was particularly interested in conceptualizations of the 4th dimension, and was the first to use the term "tesseract" to describe the four-dimensional cube. Hinton argued that people could develop perceptions of higher dimensions if they rid themselves of the conventions of right and left, up and down, through a process of "casting out the self" (De Witt, 2013). In *The fourth dimension*, Hinton (1904) describes how our sensory habits and our capacity to make sense of the world in three dimensions can be altered and opened onto a fourth dimension. Like many others in the late nineteenth century, in response to developments in non-Euclidean geometry and

E. de Freitas (✉)
Education and Social Research Institute, Manchester Metropolitan University, Manchester, UK
e-mail: L.de-freitas@mmu.ac.uk

E. Faggiano et al. (eds.), *Innovation and Technology Enhancing Mathematics Education*, Mathematics Education in the Digital Era 9,
DOI 10.1007/978-3-319-61488-5_11

topology, there was a widespread interest in rethinking human spatial sense. Along with Charles Hinton, others working in England, such as Charles Dodgson (Lewis Carroll) and Edwin Abbott, the author of Flatland (1884/2007), pursued this burgeoning interest in experiments that contest the limits of human perception.

By examining the experiential qualities in one, two and three dimensions, Charles Hinton attempted to generalize a method or model for how the human mind might come to understand a four-dimensional space. Hinton was motivated to consider the implications of this work for mathematics teaching and learning. Speculating philosophically about these implications led Hinton to propose a "higher" form of existence whereby one might tap into and perceive the fourth dimension. Rather than looking to mystical and religious sources for apprehending this higher being, he turned to mathematics and the physical sciences to study the "physical reality of the fourth dimension" (Hinton, 1902, p. 142). Through developing a more advanced spatial sense, Hinton aspired to what he called the "higher man".

> We must learn to realize the shapes of objects in this world of the higher man; we must become familiar with the movements that objects make in his world, so that we can learn something about his daily experience, his thoughts of material objects, his machinery (Hinton, 1904, p. 121).

In this chapter, I describe Hinton's methods for developing perception of a 4th *geometric* dimension. I show how his approach can be characterized as a kind of materialist approach to the study of mathematics because of the way it fuses mathematics with matter. I then discuss recent attempts to build digital mazes that develop spatial sense in 4 dimensions, and show how these are directly linked to Hinton's ideas. I focus on how colour and movement in digital environments are used to develop a distinctive kind of spatial sense. Digital environments allow for powerful uses of colour due to the pixel screen and the simulation of depth and movement. This chapter sheds light on innovative uses of digital software for developing student spatial sense, and thereby responds to a need to think more broadly about spatial sense in mathematics education (Kinache, 2012; Wai, Lubinski, Benbow, 2009). I discuss how 4-D digital mazes work and how their use of colour is linked to the philosophical insights of Charles Hinton. My aim is to contribute to discussions about the nature of spatial sense and spatial reasoning, and to point to possible directions for future research on inventive approaches to geometry. In the first sections, I discuss the insights of Hinton, and in the last sections I discuss two examples of 4-D digital mazes, as well as findings regarding participant skill at nagivating such mazes.

2 A New Materialist Approach to Space

Thought experiments run throughout Hinton's work. In these thought experiments, he melds mathematics and physics in creative speculative ways. In *Scientific Romances* (1884) he asks that we imagine a skew line moving through a plane and then states "If the plane were of such a nature as to close up behind the line, if it were of the nature of a fluid, what would be observed would be a moving point." (Hinton, 1884, p. 13). This focus on a *fluid* plane on which and through which points, lines, planes and other entities might pass as *physical-mathematical objects*, leaving various traces at varying speeds, allows him to speculate about the nature of both geometry and perception. He carries through with this thought experiment, rigorously pursuing the implications in terms of geometric space, and describes an entire structure of enmeshed lines—a "framework"—cutting through this fluid plane at different angles, and thus producing—for those who live and perceive *on the plane*—points that move across the plane at varying speeds. Indeed, the permanence of any static point on the plane will be the *effect* of one such perpendicular line moving through the plane, and have no other presence than that which is sustained through a particular kind of movement. In this way, Hinton centers movement in the making of mathematics, showing how any individuated bodies (like points) are secondary effects of a particular kind of movement. It's important to note that this secondary effect is not a Platonic image or reflection of a hidden world, because the movement that engenders the form does not operate according to resemblance or copy. Hinton does not hesitate to move back and forth between mathematical entities and material ones as he pursues these ideas:

> Let us now assume that instead of lines, very thin threads were attached to the framework: they on passing through the fluid plane would give rise to very small spots. Let us call the spots atoms, and regard them as constituting a material system in the plane (Hinton, 1884, p. 14).

The threads are woven together and form connected shapes that pass through the fluid plane, creating an effect that lasts for some period of time: "These moving figures in the plane are but the traces of the shapes of threads as those shapes pass on. These moving figures may be conceived to have a life and a consciousness of their own." (Hinton, 1884, p. 15). Like many other philosophers in the 1890s, Hinton is exploring the mixture of matter and mind, foreshadowing the convictions of contemporary theorists who describe themselves as new materialist (Coole & Frost, 2010). In *A New Era of Thought* (1888), Hinton says that he prefers to use the term "higher matter" rather than "higher space" because it does not make sense to split concrete matter into extension and impenetrability (Hinton, 1888, p. 106). He hopes that "the horizon of thought is altered" (Hinton, 1888, p. 107), not away from matter and towards spiritual existences, but towards the enhanced apprehension of the material. He will suggest that apprehending such "higher matter" demands an attention to detail—a proliferation of detail—so that awareness comes from saturation of detail rather than from generalizing from cases. In other words, He is seeking a way of developing spatial sense that is more immersive, and less based on ideas of abstraction whereby a concept transcends the details and the examples in which it is said to be instantiated. This is an important point as we turn to questions of learning *how to* perceive in four geometric dimensions. Immersive experiences are detail-oriented methods of learning, in which the environment is saturated with pedagogic potential. And although he seeks to study our experiences in three dimensions to learn about the fourth (which might be deemed a kind of generalization), his method of apprehension will ultimately depend on maximizing the apprehension of details.

These highly speculative claims point to Hinton's interest in mapping the virtual dimension of space. He links consciousness to space in ways that resonate with the historian of mathematics Gilles Chatelet's work on the virtual dimension of matter (2000), asking "Can our consciousness be supposed to deal with a *spatial profile of some higher actuality*?" (Hinton, 1884, p. 16. my italics). Moreover, the movement that we perceive is but the trace of a "higher" movement, a movement that produces images that *do not resemble* the threaded structures exactly, but are in direct contiguous and haptic relation (an indexical relation) to those structures. What makes this approach highly important for experiments today regarding perception and four dimensions is Hinton's attempt to think four dimensions according to these kinds of thought experiments where selected invariants—like movement and force—are carried over to the other dimension, although modulated due to the ways that sensation always involves more than geometric relations. Hinton, like others working today on digital 4-D mazes, stresses that our experience in three dimensions can be examined and used to develop a perception in four dimensions.

> Again, it sometimes appears to be thought that the fourth dimension is in some way different from the three which we know. But there is nothing mysterious at all about it. It is just an ordinary dimension tilted up in some way, which with our bodily organs we cannot point to." (Hinton, 1884, p. 46).

We hear in this citation how the fourth dimension is just "tilted up" in some way that we are unable to perceive. But he moves on to make two very important claims about how to build awareness of the fourth dimension. Two everyday spatial experiences are pivotal in his expanding to the fourth dimension. The first is the common feeling of being surrounded or bounded by a space of higher dimension, which implies that all three-dimensional objects touch and are contiguous with this higher dimensional space. The second is the feeling of the continuum, and the fact that any space must be composed of an *infinite number* of objects of lesser dimension (i.e. an infinite number of planes compose a solid). This implies that any four-dimensional space must be composed of an infinite number of three-dimensional objects.

3 Continuity and the Infinitely Smooth Texture of Matter

> A being existed in four dimensions must then be thought to be as completely bounded in all four directions as we are in three. All that we can say in regard to the possibility of such beings is, that we have no experience of motion in four dimensions. The powers of such beings and their experience would be ampler, but there would be no fundamental difference in the laws of force and motion. (Hinton, 1884, p. 17)

And yet we will see when we discuss digital four dimensional mazes that forces such as gravity will *by necessity* act differently when perceived in four dimensions. But Hinton's point is that in order to become conscious of how these differences will be lived in the higher dimension, we must attend to the "infinitely minute" in matter—"the ultimate particles of matter"—because only then might we be able to compare magnitudes in all four dimensions (Hinton, 1884, p. 21). When we can perceive the infinitely minute, we can then begin to operate according to proportions between such magnitudes in some pragmatic sense (Hinton, 1884, p. 21). It is interesting to note that Hinton seems to be advocating for a certain *infinitesimal* relationship that might help us connect with the fourth dimension. He describes the "thin" dimension of entities, lines for instance, much like Evariste Torricelli might have described the inflatable width of a line, when he broke with the Euclidean definition of line in order to develop the infinitesimal calculus. For Hinton, the same materiality and plasticity of an unperceived *geometric* dimension can be imagined as "thin", if we accept that matter quivers and vibrates with potentiality:

> The direction in which it is thin is in a direction which we do not know, in which we cannot move. But although we cannot make any movements which we can observe with our eyes in this direction, still the thin film—thin though infinitely extended in any way which we can measure—this thin film vibrates and quivers in this new direction, and the effects of its trembling and quivering are visible in the results of molecular motion. It only affects matter by its movement in directions at right angles to any paths which we can point to or observe, *and these movements are minute*: but still they are incessant, all-pervading, and the cause of movements of matter. *It is smooth—so smooth that it hinders not at all the gliding of our earth in its onward path*" (Hinton, 1884, p. 52. My italics).

This reference to the fourth direction as being infinitely smooth points again to the infinitesimal as the "smallest interval" or miniscule thread by which the dimensions are stitched together, where the stitch is so fast and the weave so tight, space becomes so smooth and so intense that we are able to slip across dimensions. This is linked to Riemann's influential essay "The hypotheses which lie at the foundation of geometry" whose ideas Hinton seems to echo at times. And yet the smoothness of Riemann's continuous manifold was—at the infinitesimal scale—a Euclidean smoothness, whereby the manifold is glued together or composed of patches of Euclidean flat planes (Plotnitsky, 2012). In the 1970s, this use of the infinitesimal developed into synthetic differential geometry and smooth infinitesimal analysis (Bell, 2014). For instance, the mathematician F.W. Lawvere used the infinitesimal to develop this kind of mathematics, assuming the continuum as an autonomous notion, not requiring the notion of the discrete, substituting the idea of the limit with the idea of the nilpotent infinitesimal, a quantity so small that some power of it vanishes. I draw these historical links between Hinton's ideas and developments in mathematics to show that his musings are not mere quackery. I also want to draw this link to the infinitesimal because it has played such a significant role in fueling the kind of thought experiments that Hinton pursues so well.

The infinitely small plays a pivotal role in how Hinton imagines we are capable of cognition in four dimensions. Hinton speculates that movement at the *quantum level* and the "minute portions of matter" may "go through four-dimensional movements and form four-dimensional structures" (Hinton, 1888, p. 109). Like Henri Bergson during the same era, along with many other continental and English philosophers, Hinton was inspired by developments in physics regarding the odd behavior of quantum particles. He suspected that these particles were capable of moving through a geometric fourth dimension. Thus our power to perceive in four dimensions may in fact involve our learning the movements of the very small particles of matter. He then suggests that *thought itself* might be considered the infinitely small, and *therefore the movement of thought is a movement through the fourth dimension.*

The goal of apprehending in four dimensions corresponds to the goal of grasping thought in all its mobility—"by observing, not what we can see, but what we can think." (Hinton, 1888, p. 110). Hinton makes a Spinoza-like attempt to relink thought *with* matter, and to ascribe to thought a particular kind of movement that is *not* representational of three-dimensional extension, *but accords to four dimensional movement*. Thought, suggests Hinton, and its "small molecules in the brain … might go through four-dimensional movements and form four-dimensional structures (Hinton, 1888, p. 110). Despite how odd such claims may sound, his approach to consciousness links up with current interest in quantum computing, where the temporal dynmanics of 'cognition' disobey the usual space-time rules. For Hinton, we are indeed four-dimensional creatures, ill-equipped to percieve beyond three dimensions, but nonetheless thought and imagination, at their freest, plug into the 4th dimension. Hinton's insights into how we might come to perceive the 4th dimension are thus linked to related projects, at the turn of the twentieth century, of developing an "intuitive method" that might tap into the virtual mobility of matter (Bergson, 1903). These projects looked for a virtual dimension buried in matter. More recently, this kind of work supports attempts to study learning in terms of *the movement of thought*, and not simply the movement of already individuated bodies (de Freitas & Ferrara, 2015; see also Chap. The Coordinated Movements of a Learning Assemblage: Secondary School Students Exploring Wii Graphing Technology).

4 Movement and Spatial Sense

Hinton was influenced by the work of Lobatchewsky and Bolyai (1830s) on alternative axiomatic foundations for geometry, as well as the work of Riemann (1850s) who had developed analytic tools for mapping the distinctive spatial characteristics of n-dimensional manifolds, and had introduced a new way of thinking about relationships between geometry and space. This latter work, in particular, had led to all sorts of new ways of defining dimension. In 1912, Poincaré (1854–1912) used an inductive strategy for defining dimension that was based on the concept of boundary and border. A space is n + 1 dimensional, suggested Poincaré, if its border is n dimensional. For instance, a 3 dimensional object has 2 dimensional faces (figures), and a 4 dimensional cube (a hypercube) has 3 dimensional faces (cubes). Another way of conceiving dimension is through movement. For instance, one might drag or move an object to enter a new dimension: by dragging a 0-dimensional object in some direction, one obtains a 1-dimensional object. By dragging a 1-dimensional object in a *new direction*, one obtains a 2-dimensional object. These diverse ways of thinking about the concepts of dimension and orientation were debated throughout the nineteenth century.

Like Poincaré, Hinton described the process of moving to higher dimensions in terms of boundaries and movement. He noted that a line is divided in two by a boundary point, that a plane is divided into two distinct planes by a boundary line, and a volume likewise by a boundary plane. In each of these cases, the next dimensional space is generated through the *new movement* of the lesser dimensional entity.

> Thus, going on, we may say that space is that which limits two portions of higher space from each other, and that our space will generate the higher space by moving in a direction not contained in itself (Hinton, 1904, p. 122)

What is exciting is the acknowledgement of a movement "in a direction not contained in itself" which underscores two important ideas: (1) this movement is *unrecognizable as movement in the original space*, and (2) such movement points to a virtual dimension or potentiality of space. Because the movement is unrecognizable as movement in the n-1 dimensional space, work on perception in higher dimensions helps us imagine how there might be a movement *in this world* that is beyond our perception.

Hinton (1904) uses examples of physical movement to develop his argument. He describes the movement of a line and a spiral through a film (Fig. 1), where the intersection of the spiral and film would be a point moving in a circle (see dotted line in Fig. 1).

He then introduces a second movement, that of the film itself moving "vertical" or perpendicular to the lateral space of the film. In the first case, any dweller on the film would simply perceive the circle being drawn on the film, but in the second case, the film dweller would develop an awareness of the film's trajectory through time. This introduces a reflexive perception, that is, a perception of one's

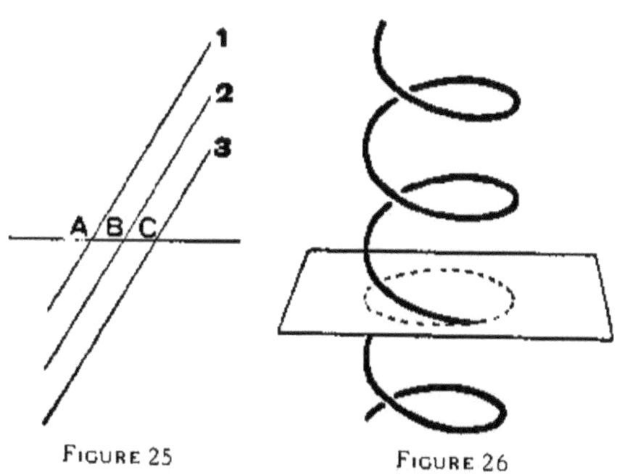

Fig. 1 Hinton's line and spiral moving through a material film

environment moving, and thus also introduces, according to Hinton, consciousness. This consciousness is distributed across both the film and the spiral, by way of their intersection:

> In the film the permanent existence of the spiral is experienced as a time series—the record of traversing the spiral is a point moving in a circle. *If now we suppose a consciousness connected with the film* in such a way that the intersection of the spiral with the film gives rise to a conscious experience, we see that we shall have in the film a point moving in a circle, conscious of its motion, knowing nothing of that real spiral the record of the successive intersections of which by the film is the motion of the point (my italics, Hinton, 1904, p. 125).

Through this combination of movements and diverse points of view, Hinton describes the "apparent" motion within the film, felt by the "plane of consciousness" or "space of consciousness" as a record of the movement—"each atom at every moment is not what it was, but a new part of that endless line which is itself." (Hinton, 1904, p. 125). These atoms are "acting, living" and partake always of at least two motions, one associated with their interactions within the film, reflecting spatial-temporal relationships immanent to that plane, and the other associated with the film itself moving in an altogether different spatio-temporal dimension. Thus the atoms can only "read off" this motion "in the film" as the trace of an altogether different kind of movement, a movement that is in some *strong sense* inconceivable within the film. This kind of leaping into another dimension where the movement is literally inconceivable in the original space is exactly what is needed to begin operating in four dimensions. Hinton's approach to the materiality of the spiral intersecting the plane lends itself to a new materialist reading of mathematics (de Freitas & Sinclair, 2014; a different example is given in Chap. Returning to Ordinality in Early Number Sense: Neurological, Technological and Pedagogical Considerations). He isn't shy to treat circles and other mathematical entities as material or physical objects, and does so in order to think differently about the way that mathematics *is in* the material world. His work is filled with these kinds of thought experiments.

> Let us now make the supposition that film after film traverses these higher structures, *that the life of the real being is read off again and again in successive waves of consciousness*. There would be a succession of lives in the different advancing planes of consciousness each differing from the preceding, and *differing in virtue of that will and activity* which in the preceding had not been devoted to the greater and apparently most significant things in life, but the minute and apparently unimportant (Hinton, 1904, p. 126. my italics)

Apprehending this motion that is inherent to life and matter, that is intrinsic to being and becoming, is the task of the "higher man". This task requires that 'man' has "a consciousness of motion which is not as the motion he can see with the eyes of the body" (Hinton, 1904, p. 128). Most notably, this requires becoming conscious of a motion that is not discernable or perceivable in our current perceptual organization. If we are to perceive the fourth dimension, we must achieve this decentering of our current perceptual apparatus. Hinton suggests that this approach to the fourth dimension reflects a certain Eastern philosophy of nature and matter. I would also argue that his approach resonates strongly with that of the French philosopher Henri

Bergson who was highly influential in the same period in which Hinton was formulating and writing his ideas. Bergson (1896) argues that there is a difference between *sensori-motor movement*, which is movement trapped by the measure conventions of the perceptible world, and the movement of the whole of duration, a movement that taps the virtual potentiality of matter.

Hinton uses the Pythagorean theorem to elaborate his approach to space. He considers an alternative world where the Pythagorean theorem states that a *sheared* square on the hypotenuse is the difference between the two squares on the other sides of the right triangle. Shearing retains the area of a square, but changes the shape. In this alternative world, we imagine that the inhabitants 'see' the square and its shear image as the same (or as equivalent) if there is a "shear rotation" movement that generates one from the other. As long as such a motion can be identified, two figures will be considered equal. In this alternative world, we emphasize that our own rules no longer apply, and the Pythagorean theorem takes on a different import. Notably, Hinton is experimenting here with how perception and geometry are entwined, and how different geometries entail different topological relationships. He is imagining a different geometry so that he might problematize the very notion of distance or metric, and at the same time link it more directly to matter: "Hence distance independent of position is inconceivable, or practically, distance is solely a property of matter." (Hinton, 1904, p. 136). This reference to "matter" marks his empirical approach, and his realization that only observation and experiment can help us decide whether *our* Pythagorean theorem or this other version of the Pythagorean theorem is appropriate: "There is nothing to connect the definition of distance with our ideas rather than with his, except the behavior of an actual piece of matter." (Hinton, 1904, p. 136).

For Hinton, alternative non-Euclidean geometries should inspire us to reconsider the nature of perception and also the limits of material agency. According to Hinton, the discoveries of non-Euclidean geometry have significance for all sorts of reasons, one of them being how they force us to consider both materiality and sensation in new ways:

> By immersing the conception of distance in matter to which it properly belongs, it promises to be of the greatest aid in analysis; for the effective distance of any two particles is the product of complex material conditions and cannot be measured by hard and fast rules. Its ultimate significance is altogether unknown. It is a cutting loose from the bonds of sense, not coincident with the recognition of a higher dimensionality, but indirectly contributory thereto (Hinton, 1904, p. 140).

5 Depth of Field and Colourism

Before turning to a discussion of how Hinton's ideas are linked to current digital four dimensional mazes, it's important to discuss some of the visual cues that are used when humans navigate through space. Visual perception in multiple dimensions entails the notion of *depth of field*. Cutting (1995, 1997) proposes that we think of space in three ways—personal space, action space, and vista space—each defined in terms of their proximity to the body. Perceiving in each of these spaces draws on different kinds of perceptual habits. He shows how we decode various depth cues when making sense of images, identifying various key aspects of unconscious perception. Such cues are crucial as we move through spaces and integrate paths, and these are used extensively by digital game designers, as they develop complex spatial mazes and puzzles. Based on Cutting (1995, 1997), I list here key cues for depth perception, as they play a crucial role in digital maze navigation:

(1) Occlusion—where part of an object is occluded behind another—is a standard way of rendering depth or dimension, although it does not convey any measure to the depth and thus offers limited information.
(2) Height in the visual field renders depth through the measures of relations among the bases of objects in the image.
(3) Relative size [linear perspective] whereby perspective (linear or otherwise) renders depth perception. This engenders a ratio between objects' positions rather than any objective measure of depth between them.
(4) Relative density refers to the projected number of similar objects or textures per solid visual angle. According to Cutting (1997) it was not until the 15th century that artists began rigorously using all four of these in conjunction. He also points out that digital media coordinate all four strategies in generating an image from a single point of view.
(5) Binocular disparity is the difference in relative position of an object as projected on the retinas of the two eyes. When disparities are small, we perceive solid space. When disparities are greater—often when an object is very near—we suffer double vision.
(6) Motion perspective where depth is rendered through a moving observer. This technique is very good for judging absolute depths, rather than just determining which objects are in front and which behind.
(7) Texture gradients.
(8) Brightness and shading.
(9) Kinetic depth concerns depth derived from the movement of parts of the image. This often entails (dis)occlusion revealed through motion. With respect to what sort of geometry applies in each of these spaces, Cutting (1997) finds that personal space is perceived as "Euclidean" but that action space is perceived as affine, although movement through such space can reconfigure it as Euclidean.

Most of these points, and especially the last regarding "kinetic depth" perception, are actively deployed in making sense of moving images. Deleuze (1989) suggests that Orson Welles is the master of depth of field in the moving image. In each of his films, the camera plunges through space, away from the viewer, deep into a *beyond* that unfolds through optical barriers, and into nested rooms and distant corners. It feels as though the camera is travelling into space, penetrating a volume of space, poking through what might have been a two-dimensional image and carving out a new dimension. In the history of cinema, depth of field is a hallmark of the first films that documented movement (Lumière and the 1895 "Exiting of the factory"), but Deleuze argues that these films capture depth *in* film or depth *in* image, but not depth *of* field or a depth *of* image. Depth of field achieves a different effect. This distinction can be found, he continues, in tracking changes in Western perspective drawing and painting. Prior to the 17th century, depth was conjured through carefully layered planes of vision, where each plane had its occupants and objects, each visually autonomous. In the 17th century we see paintings where an element of one plane refers directly to an element of another plane, where characters address each other across planes, and where the foreground comes into immediate contact with the background. This latter effect is depth *of* field. This contrast between a depth that is achieved through juxtaposition of differently sized images and characters, and a depth that is achieved through movement and engagement, can be seen in contrasting the cinema of Welles to that of Griffith's *Intolerance* where depth was produced by "a simple juxtaposition of independent shots (plans), a succession of parallel planes (plans) in the image" (Deleuze, 1989, p. 107).

Deleuze argues that this new depth of field is principally temporal, and indeed offers a direct image of time:

> In this freeing of depth which now subordinates all other dimensions we should see not only the conquest of a continuum but the temporal nature of this continuum: it is a continuity of duration which means that the unbridled depth is of time and no longer of space (Deleuze, 1989, p. 108).[1]

The temporal dimension of depth is thus extremely important for humans watching a moving image of a particular spatial arrangement. This discussion of the power of depth of field in the moving image helps us appreciate the nature of screen watching when navigating a 4-dimensional maze. How might navigating a 4-dimensional digital maze, ultimately a moving image that unfolds with kinetic depth, tap into this direct relationship with time? Especially as the focus on geometric relationships within such mazes tends to keep one focused on the measurement of space rather than more experiential immersive ways of apprehending? This is where Hinton and current software designers turn to colour.

In *A new era of thought* (1888), Hinton describes working with a system of colour cubes with students, developing their perceptions of the tesseract using different colours for different vertices, edges, and faces. In Hinton's system, colour is used as a way to visualize dimension, and each dimension is assigned a primary colour (Fig. 2). Surfaces that stretch into other dimensions have appropriately blended colours. For example, if a cube has edges coloured yellow, red and blue, then each of its faces would be orange (blend of red and yellow) or brown (blend of red and blue) or green (blend of blue and yellow). In order to extend this to the fourth dimension, introduce a new colour—say white—and use the blending to colour the various three-dimensional cross sections of the hypercube. In the fourth

[1] He cites Claudel who said of Rembrandt that depth was "an invitation to recall". He also refers to how Bergson and Merleau-Ponty showed how depth was principally a temporal dimension.

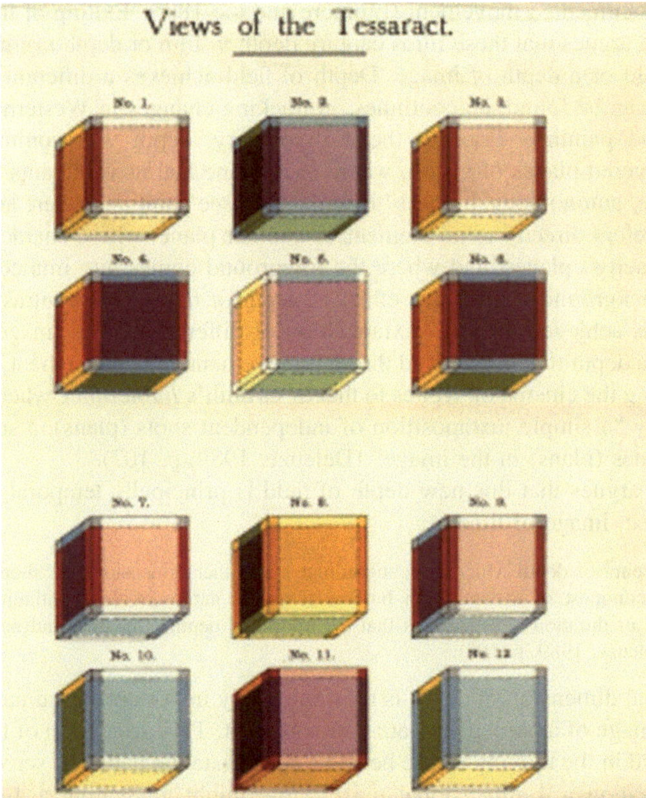

Fig. 2 Hinton's (1988) colour cubes

dimension, each cross-section is itself a cube, with its faces colour-coded appropriately, but now our vision must grapple with the blending of colours not simply on a surface but in three dimensions.

Hinton's ideas about colour and higher dimensions are fascinating for many reasons, particularly because colour is not typically studied as part of spatial reasoning and depth perception. As an under-examined quality of depth perception and orientation skills, colour blending offers important insights into the concept of dimension. Attempts to draw on the modulation of colour intensity in an image are of course familiar to artists. Deleuze (2003) suggests that Francis Bacon's skill at "colouring sensation" is what allows us to get at the distributed nature of consciousness. Bacon's eerie "scrambling and smudging" of the human figure gets to the heart of the movement of thought and what Deleuze calls the "time-image" (Deleuze, 2003, p. 127). Deleuze (2003) states "The formula for the colourists is: if you push colour to its pure internal relations (hot-cold, expansion-contraction), then you have everything" (Deleuze, 2003, p. 112). Colourism aims to show how colour itself is the variable relation or differential relation on which individuation depends. The technique of the colourist is "the production of light and even time through the unlimited activity of colour" (Deleuze, 2003, p. 112).

In pushing colour to function in this onto-generative way, Bacon and other colourists, force us to encounter the image differently. In attending to the modulated intensity of colour sensation, the eye is no longer the usual optic device, looking for resemblance, *looking for the line*, but becomes haptic and sculptural. In other words, the eye touches the image, and the sense of sight behaves like the sense of touch. In terms of developing spatial sense, this means that modulations of colour might play an important role. There needs to be more research on how modulating colour is a part of spatial reasoning. As we see below, colour is used for both representing different dimensions, but also for capturing the intensity of a fourth dimension that is folded into the first three in digital environments. In the examples I discuss, colour is used to also refer to the varying 'temperature' (hot/cold) of the fourth dimension, so as to evoke for the participants that sense of intensive quantity that infuses the material world.

6 Four-Dimensional Digital Mazes

During the last decade, perception scientists have used experiments with digital mazes to study the potential of developing human perception in 4 dimensions. In this section, I explore the way that innovative colour digital mazes draw on Hinton's earlier ideas to develop student spatial sense. Various software developers use colour to simulate the experience of moving through a fourth dimension. For instance, the mathematician Weeks (2016) has developed a game that involves moving an object (white ball) along a series of coloured paths to reach a target, such that one must navigate four dimensions (Fig. 3).[2] One can train oneself on these mazes, becoming familiar with how the colour coding and the particular corner rainbow transformations embody movement into a fourth dimension with reference to the movements one has already made. In this software, you occupy a 'bird's eye view' on the maze, and can rotate the maze as you try and decide where to move. By rotating the maze, the occluded links between pathways become visible.

One can also modify the image with a 4-D shear, so as to see where the linked tubes do not actually connect. This innovation helps one realize how the 3-D version of the maze is also a kind of occlusion, blocking the entire fourth dimension from view (see Fig. 4 to see maze without shear on left, and the same maze with shear on right). In other words, colour is used to introduce another fold in the maze, one that would be invisible to anyone operating only in 3-D space.

---

[2]Available at http://www.geometrygames.org/Maze4D/index.html.

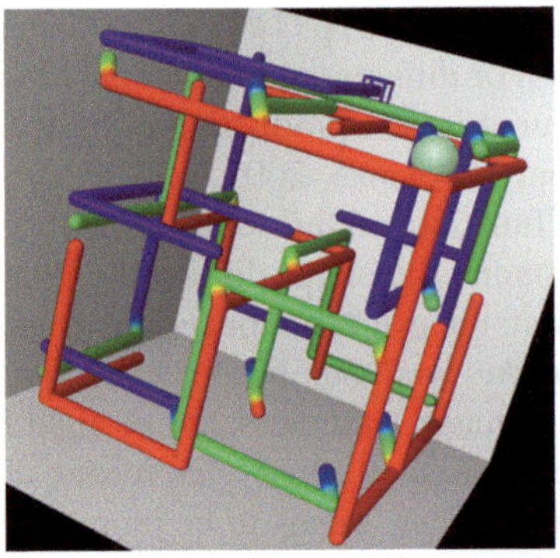

Fig. 3 Certain corners are rainbow transitions into the fourth dimension

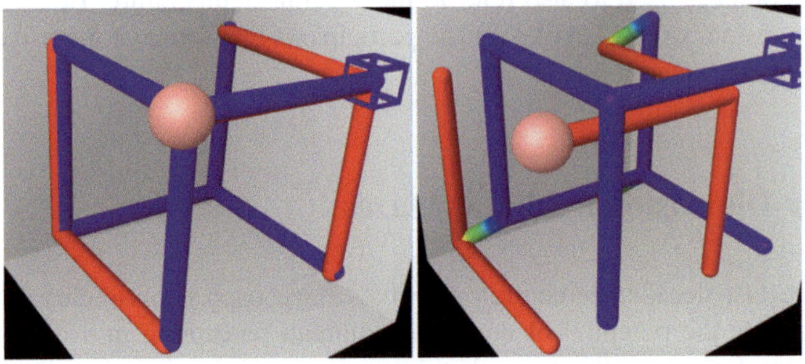

Fig. 4 The same maze, showing the 4-D shear on right

Recent attempts using digital colour mazes to develop students' perceptions of the 4th dimension show how colour can enhance "path integration" and successful spatial navigation through 4 dimensional mazes (Aflolo & Graziano, 2008). Gamers using computer simulations claim to develop competence in 4-D navigation using such visual cues (Seyranian, Colantoni, & D'Zmura, 1999). This work points to how students may be able to develop 4-D spatial maps and operate through visualizations of 4-D environments.

Aflalo and Graziano (2008) designed an innovative computer-generated maze that moved through four spatial dimensions, but was different from Weeks' maze in that there was no bird's eye view. The participants were able to develop skills in perceiving in the fourth dimension, based on their ability to perform path integration, which is a standard test in such experiments for measuring spatial sense. After moving along a winding corridor, the participants then had to 'point' back at the occluded starting point of their movement. As the authors note, these experiments indicate the potentiality of the body: "One interpretation is that the brain substrate for spatial navigation is not a built-in map of the 3-dimensional world. Instead it may be better described as a set of general rules for manipulating spatial information that can be applied with practice to a diversity of spatial frameworks" (Aflalo & Graziano, 2008, p. 1066). Much of the work on path integration shows that humans are often better when they do not rely on visual cues (Easton & Sholl, 1995; Farrell & Thomson, 1998; May & Klatzky, 2000; Presson & Montello, 1994). We often have that typical experience of getting lost in a building, and in order to proceed we look up or close one's eyes, as we try to recall how our body moved through the building. Phenomenologists have theorized this in terms of the body and its inherent capacities to navigate through space. Rush (2009) suggests that the body is thus "something that I move with, not something I move, i.e., it has the characteristic of direct motility—I do not have to place my body in order to move it." (Rush, 2009, p. 18). This observation is important as developers try to innovate with digital maze technology. Our sense of our own body's movement needs to be considered as we develop innovative technologies. Clearly path integration is only part of the story.

Research into spatial cognition in humans and other animals often uses this skill of path integration, the "short cut" test, to study spatial skills (Biegler, 2000; Newcombe & Huttenlochner, 2000; Wehner, Michel, & Antonsen, 1996). Path integration is the skill of keeping track of the various movements you've taken, summing distances and turns, so that you know where you are in relation to your starting point. Desert ants, for instance, are renowned for wandering around while scavenging, and then able to return to their starting point along a more direct route. Whether this is pheromonal or geometric knowledge (or a combination of both) is unknown.

Four-dimensional digital mazes *without* birds-eye view are built so that the fourth dimension is simulated through engagement with the digital environment. Typically, this entails building a maze so that a fourth dimension is orthogonal to the three dimensions of 3-space. In other words, the movement through the virtual maze entails a movement that is orthogonal to all the movements in 3-space. The virtual environment simulates the experience of moving in this new dimension. Aflolo and Graziano (2008) for instance, created a program which first displayed a menu of selections of either 2-D, 3-D or 4-D mazes. The program contained 100 possible examples of mazes of each type, and would automatically store the time in each maze and the angular accuracy of the subject's response at the end of the maze (the response to the path integration test. See Fig. 5e–d). The number of turns and the length of the corridors in each maze varied randomly, with given constraints, such as each corridor was between 3 to 6 units, and could not intersect itself. The view in each maze was of a virtual corridor along which the subject was moving, displaying that which was in front of the participant (Fig. 5a).

Using different commands, the subject could simulate travelling and looking in different directions within the virtual world of the maze, translating forward or backward along the corridor and rotating left or right. Texture on the walls was used to enhance perspective cues, but depth was depicted using standard perspective,

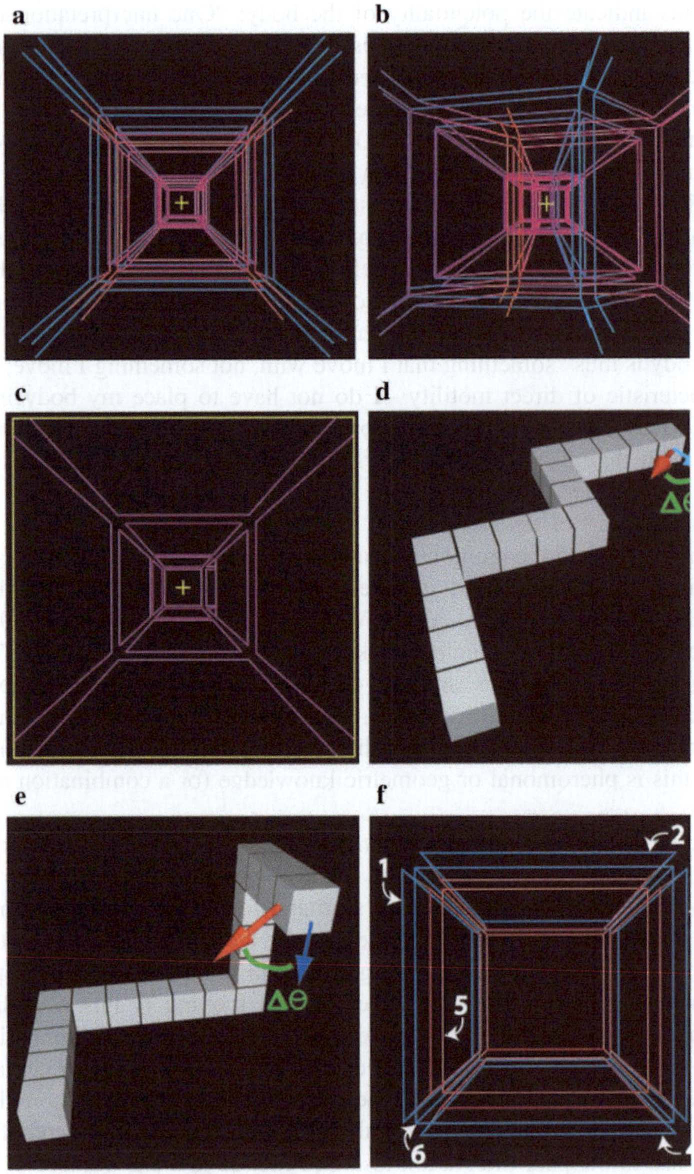

Fig. 5 Aflolo and Graziano (2008, p. 1065)

motion parallax and occlusion. In the 2-D mazes, the bends in the corridor were either to the left or right only. In the 3-D mazes, there were turns that were orthogonal to the horizontal plane. Colour was used to help them identify both the start and the end of each maze—the start cube had silver walls and the end of a corridor cube had five golden walls. Subjects never saw an outside view, and at the end of their journey, were asked to point in the direction of their starting cube, hidden behind the occluding corridor walls. A score indicating their accuracy (angular deviation from direct line joining start to finish) was revealed to the participant after each guess.

In the 4-D mazes, the cubes composing the corridors became hypercubes, and the corridor bends were right-left, up-down bend, and *hot-cold*. The terms hot and cold were used to designate directions in the mathematically defined fourth dimension, but depended on colour perception and sense of orientation. As in the case of representing 3-D space on a flat computer screen, the display of a 4-D maze relies on projection, as well as visual cues. In the virtual computer world, virtual 4-D objects are projected onto the flat display. The subject must reconstruct the 4-D object using perspective, shading, motion parallax, and occlusion—no easy task! As in the maze software of Weeks discussed above, colour was used as an additional perspective cue to aid participants. A greater degree of red hue indicated that the object (in this case a wall or edge) was more in the 'hot' direction in relation to the viewer, while a greater degree of blue hue indicated that the object was more in the cold direction in relation to the viewer. If an object was purple, then it was at the same temperature as the viewer—that is, purple indicated that it was on the same plane as the viewer in the four-dimensional space.

Visual cues and sense of orientation are strongly linked in the experience of the maze. In other words, dimensions are always determined in relation to the orientation of the viewer. Thus the viewer has a particular orientation—a front, back, top, down, hot and cold—and as the viewer rotates 90°, these relational terms (top/down) are altered *in relation to* the surrounding environment. For instance, imagine you start the maze and are looking ahead into a corridor, and then you rotate on the horizontal plane, so that the corridor is now to your left (along the left-right axis). Next, you can rotate orthogonal to that horizontal plane, so that the corridor is now above you. In other words, as you move and change your orientation, the environment occupies different dimensions (what was cold is now far away). In the 4-D topology of the space, the participant gains additional rotational degrees of freedom, labeled R4, R5 and R6. These are modeled on conventional rotational movements in 3-D space, but these vary according to the hot-cold dimension. If the viewer comes to a wall in the corridor, she now has additional rotational moves that re-orient her in relation to the four dimensions she faces. The point here is that all dimensions become rotationally swapped as the participant moves through the space. Each rotation redefines what is up/down or right/left or hot/cold. The participants only ever translation movement is in the forward/backward direction, while all other dimensions are experienced through rotation. The dimensions are thus strongly entangled or mixed through rotation, creating an unusual spatial experience. When we move up in an elevator, we do not remix or exchange our relation to the dimensions of the ground floor where we were standing before getting into the elevator. Hence, this 4-D maze invites a radically different way of orienting oneself, compared to the bird's eye view mazes discussed above. Navigating in this space entails understanding the rotational interactions of the dimensions, and seems to entail what Hinton advocated for—the need for "casting out the self" in a process of productive disorientation.

But what is this hot-cold dimension? Aflolo and Graziano (2008) describe it as a kind of additional material quality of the walls in the maze. Each wall has width, length and temperature. Generalizing from two and three dimensions, where the addition of another dimension allows escape from the previous space. For instance, an ant trapped inside a square on a plane, can escape when it is allowed to travel in three dimensions. Trying to keep the ant trapped involves layering squares on top of squares (surrounding him) and giving the initial square height so that the ant cannot crawl out (assuming it movement is on the plane). If we imagine someone trapped inside a cube, they might be also able to escape by moving in a higher-dimensional space. Suppose this dimension is temperature, so that the cube has a particular cold temperature. They can move into a higher temperature and, in theory, escape the particular trappings of the initial cube. In order to keep them trapped, one would have to add additional cubes of higher temperature, so that their movement was still contained by the stacked hypercube environment, an environment of cubes with infinite varying temperature.

Temperature is a useful term because it refers to an intensive quality that applies to all material, and brings a material perspective to the colour coding. In terms of visualizing the 4-D maze, we need to imagine the *barriers* of the corridor no longer are walls, as we imagine them, but cubes. Each cube extends top-down, right-left, hot-cold. These barriers hem the participant in, so that movement within the maze environment only ever happens in 'one' direction. In Fig. 5f, there are 6 cubes that hem the movement of the participant, acting as barriers. These stop the movement in the left-right direction, the up-down direction, and the hot-cold direction, leaving only the forward-backward direction open for movement. Cube number 1, for instance, is the left hand barrier, with one cold face, one hot face, top and bottom faces, and one near face and one further face. The other barrier cubes (right, and top and bottom) have the same construction. The red cube and the blue cube are the barriers in the temperature direction. Their absence would mean that one could move in that direction (purple). Their presence is thus a barrier to moving—just as in the two-dimensional world of the ant on the plane, a barrier is only a barrier because it occupies that plane. The actual view of the maze navigator is shown in Figure 5a, b. To simulate the depth of the maze, the structure of 1-D is repeated into the distant corridor, shrinking in size—using basic perspectival technique. In addition, the blue-red temperature dimension is shown to vary, and in the distance (depth of field) the difference between blue and red diminishes. When the participant rotates, the view becomes radically more complicated (see Fig. 5b), as does the task of managing all this visual information.

One of the fascinating Hinton-like insights of this digital maze experiment, is that if one navigated in this maze using only three dimensions, one has a "better than even-chance" odds of guessing the correct location of the original start when asked to point in that direction. This is simply because if the participant is ignoring one of four dimensions, and if they are expert at three, then their chances are pretty good. The authors used a simulated participant with perfect 3-D path integration skills, and found that such a participant pointed with an angular error of 0 Degrees and a mean accuracy of 28 Degrees. The human participants in this study, however, did seem to develop better accuracy, which implies definite improvement of skill at 4-dimensional path integration, rather than simply using 3-D skills and ignoring the fourth dimensional cues.

7 Concluding Comments

In this chapter I have explored the details of Charles Hinton's proposal for how perception in four dimensions is possible. I've shown how his ideas were linked to other ideas and techniques of observation (the moving image) that emerged during the same historical period. In particular, Hinton's ideas are linked to philosophical interest in developing an "intuitive method" as articulated by Henri Bergson, and also linked to developments in non-Euclidean geometry and topological thinking, all of which inform the theoretical turn to new materialism more recently. Although developments in non-Euclidean geometry are often presented as more abstracted from the material world, I have showed how Hinton focused on the specific ways in which geometric concepts are inherently material, and engendered through depth of field, movement, colour, and intensity (temperature).

Hinton's revolutionary approach to spatial sense is now being actualized in innovative uses of digital technology. This chapter shows how two very different attempts to build a digital 4-dimensional maze deploy many of Hinton's ideas about spatial sense, colour, movement and intensity. Moreover, these experiments in virtual navigation raise important research questions about how we can use technology to expand our ways of perceiving and being in the world, suggesting that body syntonicity is radically different in four dimensions. More research is needed on digital technology's potential for advancing our perceptual skills, and on experiments for developing our grasp of what Hinton called a "higher matter". Not only would such research serve perception studies, it would also contribute to our understanding of the nature of the relationship between geometry and the material world, and enhance students' engagement with geometric concepts.

References

Aflolo, T. N., & Graziano, M. S. A. (2008). Four dimensional spatial reasoning in humans. *Journal of Experimental Psychology: Human Perception and Performance, 34*(5), 1066–1077.

Bell, J. L. (2014). *Continuity and infinitesimals. Stanford Encyclopedia of Philosophy*. Available at http://plato.stanford.edu/entries/continuity/

Bergson, H. (1903/1999). *Introduction á la métaphysique*. (T. E. Hulme: An Introduction to Metaphysics, Trans.). Indianapolis, IN: Hackett Publishing Company, 1999.

Bergson, H. (1896/1988). *Matie`re et me'moire*. (N. M. Paul & W. S. Palmer: Matter and Memory, Trans.). New York, NY: Zone Books, 1988.

Biegler, R. (2000). Possible use of path integration in animal navigation. *Animal Learning and Behavior, 28,* 257–277.

Châtelet, G. (1993/2000). *Les enjeux du mobile.* (R. Shore & M. Zagha: Figuring space: Philosophy, mathematics and physics, Trans.). Dordrecht, The Netherlands: Kluwer Academy Press, 2000.

Coole, D., & Frost, S. (2010). *New materialism: Ontology, agency and politics.* Durham, NC: Duke University Press.

Cutting, J. E. (1997). How the eye measures reality and virtual reality. *Behavior research methods, instruments and computers, 29*(1), 27–36.

Cutting, J. E., Vishton, P. M., & Braren, P. A. (1995). How we avoid collisions with stationary and moving obstacles. *Psychological Review, 102*(4), 627–651.

de Freitas, E., & Sinclair, N. (2014). *Mathematics and the body: Material entanglements in the classroom.* Cambridge: Cambridge University Press.

de Freitas, E., & Ferrara, F. (2015). Movement, memory, and mathematics: Henri Bergson and the ontology of learning. *Studies in Philosophy of Education, 33*(6), 565–585.

Deleuze, G. (1989). *Cinema 2: The time image* (H. Tomlinson and R. Galeta, Trans.). Minneapolis, MN: The Athlone Press.

Deleuze, G. (2003). *Francis Bacon: The logic of sensation.* (D. Smith, Trans.). Minneapolis: University of Minnesota Press.

De Witt, A. (2013). *Moral authority, men of science, and the Victorian novel* (p. 173), Cambridge: Cambridge University Press. ISBN 1107036178.

Easton, R. D., & Sholl, M. J. (1995). Object-array structure, frames of reference, and retrieval of spatial knowledge. *Journal of Experimental Psychology. Learning, Memory, and Cognition, 21,* 483–500.

Farrell, M. J., & Thomson, J. A. (1998). Automatic spatial updating during locomotion without vision. *Quarterly Journal of Experimental Psychology: Human Experimental Psychology, 51* (A), 637–654.

Hinton, C. H. (1902). *The recognition of the fourth dimension. Bulletin of the philosophical society of Washington* (Vol. #14). Reprinted in *Speculations on the fourth dimension: Selected writings of Charles H. Hinton.* Ed. R.V.B. Rucker. (1980). New York: Dover Publications, pp. 142–162.

Hinton, C. H. (1904). *The fourth dimension.* Reprinted in *Speculations on the fourth dimension: Selected writings of Charles H. Hinton.* Ed. R.V.B. Rucker. (1980). New York: Dover Publications, pp. 120–141.

Hinton, C. H. (1888). *A new era of thought.* Reprinted in *Speculations on the fourth dimension: Selected writings of Charles H. Hinton.* Ed. R.V.B. Rucker. (1980). New York: Dover Publications, pp. 106–119.

Hinton, C. H. (1884). *Scientific Romances* (vol. 1). Reprinted in *Speculations on the fourth dimension: Selected writings of Charles H. Hinton.* Ed. R.V.B. Rucker. (1980). New York: Dover Publications, pp. 1–66

Kinache, B. M. (2012). Fostering spatial sense vs. metric understanding in geometry. *The mathematics teacher, 105*(7), 534–540.

May, M., & Klatzky, R. L. (2000). Path integration while ignoring irrelevant movement. *Journal of Experimental Psychology. Learning, Memory, and Cognition, 26,* 169–186.

Newcombe, N. S., & Huttenlochner, J. (2000). *Making space: The development of spatial representations and reasoning.* Cambridge, MA: MIT Press.

Plotnitsky, A. (2012). Adventures of the diagonal: Non-euclidean mathematics and narrative. In A. Doxiadis & B. Mazur (Eds.), *Circles disturbed: The interplay of mathematics and narrative* (pp. 407–446). Princeton, New Jersey: Princeton University Press.

Presson, C. C., & Montello, D. R. (1994). Updating after rotational and translational body movements: Coordinate structure of perspective space. *Perception, 23,* 1447–1455.

Rush, F. (2009). *On architecture.* New York: Routledge.

Seyranian, G., Colantoni, P., & D'Zmura, M. (1999). Navigation in environments with four spatial dimensions. *Perception,* 28 (Suppl. 7).

Wai, J., Lubinski, D., & Benbow, C. P. (2009). Spatial ability for STEM domains: Aligning over 50 years of cumulative psychological knowledge solidifies its importance. *Journal of Educational Psychology, 101*(4), 817–835.

Weeks, J. (2016). *4D mazes*. Available at http://www.geometrygames.org/Maze4D/index.html

Wehner, R., Michel, B., & Antonsen, P. (1996). Visual navigation in insects: Coupling of egocentric and geocentric information. *Journal of Experimental Biology, 199,* 129–140.

Erratum to: Introduction: Innovative Spaces for Mathematics Education with Technology

Francesca Ferrara, Eleonora Faggiano and Antonella Montone

Erratum to:
Chapter "Introduction: Innovative Spaces for Mathematics Education with Technology" in: E. Faggiano et al. (eds.), *Innovation and Technology Enhancing Mathematics Education*, Mathematics Education in the Digital Era 9, https://doi.org/10.1007/978-3-319-61488-5_1

The original version of the book was inadvertently published with incorrect text "(cf. Chap. The Duo "Pascaline and e-Pascaline": An Example of Using Material and Digital Artefacts at Primary School in de Freitas & Sinclair, 2014)" in page 2, row 9 of "Introduction: Innovative Spaces for Mathematics Education with Technology", which has to be changed as "(cf. chapter six in de Freitas & Sinclair, 2014)". The erratum book has been updated with the change.

The updated online version of this chapter can be found at https://doi.org/10.1007/978-3-319-61488-5_1

E. Faggiano et al. (eds.), *Innovation and Technology Enhancing Mathematics Education*, Mathematics Education in the Digital Era 9,
https://doi.org/10.1007/978-3-319-61488-5_12

MIX
Papier aus verantwortungsvollen Quellen
Paper from responsible sources
FSC® C105338

If you have any concerns about our products,
you can contact us on
ProductSafety@springernature.com

In case Publisher is established outside the EU,
the EU authorized representative is:
Springer Nature Customer Service Center GmbH
Europaplatz 3, 69115 Heidelberg, Germany

Printed by Libri Plureos GmbH
in Hamburg, Germany